流体力学基础

Essence of Fluid Mechanics

（英汉对照版）

〔印〕艾赛拉占·拉萨克里斯南　李松晶　编著
Ethirajan Rathakrishnan　Li Songjing

包　钢　主审
Bao Gang

科学出版社

北　京

内 容 简 介

本书通过物理描述、理论分析和具体应用求解相结合的逻辑方式,以浅显易懂的语言全面介绍了流体力学的基础知识,包括流体力学的基本概念、流体的性质、流体静力学、流体运动学和动力学、流体流动的基本定律和基本方程以及边界层理论等内容,着重分析了黏性流动中的阻力、湍流、管道流动和圆柱绕流等几个流体动力学问题。本书在通俗地论述流体力学基本理论的基础上,结合各章中给出的大量计算实例和练习题,对流体力学基本概念和理论进行了理论联系实际的阐述。

本书可用作初级流体力学课程的教材,适用于机械、航空航天、能源动力、化工、水利等专业本科生和研究生;也可用作流体力学学习和实际流体力学问题求解的入门级参考书,适用于科研人员、工程师和流体力学爱好者等。

图书在版编目(CIP)数据

流体力学基础 ＝ Essence of Fluid Mechanics：英汉对照/（印）艾赛拉占·拉萨克里斯南（Ethirajan Rathakrishnan），李松晶编著. —北京：科学出版社，2022.8

ISBN 978-7-03-072803-6

Ⅰ. ①流… Ⅱ. ①艾… ②李… Ⅲ. ①流体力学-英、汉 Ⅳ. ①O35

中国版本图书馆 CIP 数据核字(2022)第 140889 号

责任编辑：姜 红 常友丽 / 责任校对：王萌萌
责任印制：苏铁锁 / 封面设计：无极书装

科学出版社 出版
北京东黄城根北街 16 号
邮政编码：100717
http://www.sciencep.com

北京凌奇印刷有限责任公司 印刷
科学出版社发行 各地新华书店经销

*

2022 年 8 月第 一 版　开本：787×1092　1/16
2022 年 8 月第一次印刷　印张：14 1/2
字数：390 000

POD 定价：59.00元
(如有印装质量问题,我社负责调换)

Preface

One of the primary objectives of this book is to make the readers feel comfortable with the concise form of the book. The entire spectrum of the fluid mechanics is presented in this book, with necessary explanations on every aspect. The materials covered in this book are so designed that any beginner can follow them comfortably.

The terminologies of the subject, the units and dimensions are discussed in Chapter 1. The properties of the fluids are given in Chapter 2 to make the foundation for the understanding of fluid flow. The essence of fluid mechanics including the fluid statics, fluid kinematics and dynamics, being the vital part for the subject of fluid flow, is given in Chapter 3 and Chapter 4. Several interesting problems, such as viscous flow, drag of bodies turbulence, pipe flow and flow passing cylinders, in fluid dynamics are covered and are treated at the undergraduate level in Chapter 5. The content of boundary layer is also included as the sixth chapter to give readers a complete sense of fluid mechanics. A considerable number of solved examples are given in all chapters to fix the concepts introduced and a large number of exercise problems along with answers are listed at the end of these chapters to test the understanding of the material studied.

We would like to express our sincere thanks to Professor Kojiro Suzuki, Department of Advanced Energy, Graduate School of Frontier Sciences, University of Tokyo, Japan, for his suggestion to bring this as a book. Indeed this first impression of Professor Suzuki, on the material in the form of lecture notes for a series of lectures Professor Ethirajan Rathakrishnan gave to the graduate students of the School of Frontier Sciences, at the Kashiwa Campus of University of Tokyo, in the year 2011, is the prime factor for the development of this book.

Our sincere thanks will be given to Dr. Yasumasa Watanabe, Department of Aeronautics and Astronautics, University of Tokyo, Japan, for his suggestions and help during the entire course of the development of the manuscript of this book. The Moody's diagram in this book was created by Dr. Watanabe. Professor Bao Gang from Harbin Institute of Technology, China, carefully reviewed the whole book manuscript and we are very grateful for his systematical and helpful advice to rearrange the structure of the manuscript.

We thank our doctoral students S. M. Aravindh Kumar and Anuj Bajpai from Indian Institute of Technology Kanpur for taking part in the writing and checking of the English part of the manuscript. We also give our appreciation to our doctoral students Zhang Shengzhuo, Lü Xinbei, Jia Weiliang and Zhu Yunfeng, Master degree students Cai Shen, Yuan Shuai, Wu Haicheng, Zhai Hui and undergraduate student Lü Hongbin from Harbin Institute of Technology, for preparing some parts of the chapters in Chinese, checking the solutions manual, doing formatting and editing, and giving some useful suggestions.

We sincerely thank our undergraduate and graduate students in India, China and other countries, who are directly and indirectly responsible for the development of this book.

The financial support extended by the National Natural Science Foundation of China (No. 51675119) and the Continuing Education Centre of the Indian Institute of Technology Kanpur, for the preparation of the manuscript of this book is gratefully acknowledged.

Ethirajan Rathakrishnan and Li Songjing
March 2021

前　言

作者撰写本书的首要目标之一是让读者对本书的简洁形式感到轻松舒适。本书给出了流体力学各方面内容的必要解释，从而介绍了流体力学的整体脉络。在本书所涵盖内容的选取上，我们力求让任何初学者都能够很容易地掌握所有内容。

本书第 1 章给出流体力学的基本术语、单位和量纲。第 2 章介绍流体的性质，为后续理解流体流动的基本理论奠定了基础。第 3 章和第 4 章阐述流体力学的基本理论，包括流体静力学、流体运动学和动力学，它们是流体流动学科的重要组成部分。第 5 章介绍几个普遍被关注的流体动力学问题，如黏性流动中的阻力、湍流、管道流动和圆柱绕流等。第 6 章主要介绍边界层的相关内容，这部分内容设置能够使读者对流体力学的认知更加完整。本书所有章节都给出了很多求解实例以进一步加深读者对所学内容的掌握，并在各章末尾列出了大量的练习题和答案，以测试读者对所学内容的理解。

我们衷心感谢日本东京大学前沿科学研究生院先进能源系铃木小次郎教授。2011 年铃木教授聆听了 Ethirajan Rathakrishnan 教授在东京大学柏校区为前沿科学研究生院学生开设的流体力学系列讲座，对当时 Ethirajan Rathakrishnan 教授的授课讲义印象非常深刻，提出了将授课讲义整理并出版的建议。这就是本书的由来。

我们由衷地感谢日本东京大学航空航天系渡边泰正博士在本书手稿构思的整个过程中给予的建议和帮助，书中的穆迪图就是渡边博士绘制的。中国哈尔滨工业大学的包钢教授对整本书稿进行了审阅，感谢包钢教授给出的对全书框架进行调整的系统而有益的建议。

感谢印度坎普尔理工学院的博士生 S. M. Aravindh Kumar 和 Anuj Bajpai，他们参与了本书英文部分的核对。感谢中国哈尔滨工业大学的博士研究生张圣卓、吕欣倍、贾伟亮、朱鋆峰，以及硕士研究生蔡申、袁帅、吴海成、翟辉和本科生吕宏彬，他们参与了本书中文部分章节的素材整理、文字修改和习题的核对工作，完成了大量格式编辑工作，并给出了一些建议。

真诚地感谢我们在印度、中国和其他国家的本科生和研究生，他们直接或间接地参与了本书的整理工作。

感谢中国国家自然科学基金（项目编号：51675119）和印度坎普尔理工学院继续教育中心为本书出版提供的资金支持。

艾赛拉占·拉萨克里斯南　李松晶

2021 年 3 月

CONTENTS

Preface

Chapter 1　Basic Concepts ··· 1

   1.1　Introduction ··· 1
   1.2　Some Basic Facts About Fluid Mechanics ··· 1
   1.3　Fluids and the Continuum ·· 5
   1.4　The Perfect Gas Equation of State ·· 7
   1.5　Regimes of Fluid Mechanics ··· 9
       1.5.1　Ideal Fluid Flow ··· 9
       1.5.2　Viscous Incompressible Flow ··· 10
       1.5.3　Gas Dynamics ·· 10
       1.5.4　Rarefied Gas Dynamics ·· 11
       1.5.5　Flow of Multicomponent Mixtures ··· 13
       1.5.6　Non-Newtonian Fluid Flow ·· 13
   1.6　Dimension and Units ·· 13
   1.7　Law of Dimensional Homogeneity ·· 14
   1.8　Summary ·· 17
   1.9　Exercises ·· 17

Chapter 2　Properties of Fluids ·· 19

   2.1　Introduction ·· 19
   2.2　Basic Properties of Fluids ··· 19
       2.2.1　Pressure of Fluids ··· 20
       2.2.2　Temperature ·· 21
       2.2.3　Density ·· 22
       2.2.4　Viscosity ·· 23
       2.2.5　Compressibility ··· 28
   2.3　Thermodynamic Properties of Fluids ·· 29
       2.3.1　Specific Heat ··· 29
       2.3.2　The Ratio of Specific Heats ··· 30
       2.3.3　Thermal Conductivity of Air ·· 31
   2.4　Surface Tension ·· 32
   2.5　Summary ·· 34
   2.6　Exercises ·· 34

流体力学基础 Essence of Fluid Mechanics

Chapter 3　Fluid Statics　36

3.1　Introduction　36
3.2　Scalar, Vector and Tensor Quantities　36
3.3　Body and Surface Forces　37
3.4　Forces in Stationary Fluids　38
3.5　Pressure Force on a Fluid Element　39
3.6　Basic Equations of Fluid Statics　40
　　3.6.1　Hydrostatic Pressure Distribution　41
　　3.6.2　Measurement of Pressures　43
　　3.6.3　Units and Scales of Pressure Measurement　46
3.7　The Atmosphere　46
　　3.7.1　The International Standard Atmosphere　47
　　3.7.2　Calculations on the Stratosphere　47
　　3.7.3　Calculations on the Troposphere　49
3.8　Hydrostatic Force on Submerged Surfaces　54
3.9　Buoyancy　56
3.10　Summary　57
3.11　Exercises　58
References　62

Chapter 4　Kinematics and Dynamics of Fluid Flow　63

4.1　Introduction　63
4.2　Description of Fluid Flow　63
　　4.2.1　Lagrangian and Eulerian Methods　63
　　4.2.2　Local and Material Rates of Change　64
　　4.2.3　Graphical Description of Fluid Motion　66
4.3　Basic and Subsidiary Laws　68
　　4.3.1　System and Control Volume　68
　　4.3.2　Integral and Differential Analysis　69
4.4　Basic Equation　69
　　4.4.1　Continuity Equation　70
　　4.4.2　Momentum Equation　70
　　4.4.3　Equation of State　72
　　4.4.4　Boundary Layer Equation　73
4.5　Rotational and Irrotational Motion　75
　　4.5.1　Circulation and Vorticity　75
　　4.5.2　Stream Function　76
　　4.5.3　Relationship Between Stream Function and Velocity Potential　77
4.6　Potential Flow　78

		4.6.1 Two-Dimensional Source and Sink	81
		4.6.2 Simple Vortex	82
		4.6.3 Source-Sink Pair	84
		4.6.4 Doublet	84
	4.7	Flow Past a Half-Body—Combination of Simple Flows	88
	4.8	Summary	97
	4.9	Exercises	97

Chapter 5 Several Problems of Fluid Dynamics · 114

	5.1	Introduction	114
	5.2	Viscous Flows	114
	5.3	Drag of Bodies	117
		5.3.1 Pressure Drag	118
		5.3.2 Skin Friction Drag	124
		5.3.3 Comparison of Drag of Various Bodies	125
	5.4	Turbulence	128
	5.5	Flow Through Pipes	136
	5.6	Flow Past a Circular Cylinder Without Circulation	142
	5.7	Flow Past a Circular Cylinder With Circulation	146
	5.8	Summary	151
	5.9	Exercises	151
	References		162

Chapter 6 Boundary Layer · 163

	6.1	Introduction	163
	6.2	Boundary Layer Development	164
	6.3	Boundary Layer Thickness	167
		6.3.1 Displacement Thickness	168
		6.3.2 Momentum Thickness	170
		6.3.3 Kinetic Energy Thickness	171
		6.3.4 Non-Dimensional Velocity Profile	172
		6.3.5 Types of Boundary Layer	173
	6.4	Boundary Layer Flow	175
	6.5	Boundary Layer Solutions	179
	6.6	Momentum-Integral Estimates	179
		6.6.1 Conservation of Linear Momentum	179
		6.6.2 Karman's Analysis of Flat Plate Boundary Layer	181
	6.7	Boundary Layer Equations in Dimensionless Form	182
	6.8	Flat Plate Boundary Layer	189

6.8.1 Laminar Flow Boundary Layer ………………………………………… 190
  6.8.2 Boundary Layer Thickness for Flat Plate ……………………………… 192
6.9 Turbulent Boundary Layer for Incompressible Flow Along a Flat Plate ………… 201
6.10 Flows With Pressure Gradient ………………………………………………… 205
6.11 Laminar Integral Theory ……………………………………………………… 206
6.12 Summary ………………………………………………………………………… 214
6.13 Exercises ………………………………………………………………………… 214
References ……………………………………………………………………………… 217

目　　录

前言

第 1 章　基本概念 ··· 1
 1.1　引言 ·· 1
 1.2　流体力学概况 ··· 1
 1.3　流体和连续介质 ··· 5
 1.4　完全气体状态方程 ··· 7
 1.5　流体力学范畴 ··· 9
  1.5.1　理想流体流动 ·· 9
  1.5.2　黏性不可压缩流动 ··································· 10
  1.5.3　气体动力学 ··· 10
  1.5.4　稀薄气体动力学 ······································ 11
  1.5.5　多元混合流动 ·· 13
  1.5.6　非牛顿流体流动 ······································ 13
 1.6　量纲和单位制 ··· 13
 1.7　量纲一致性原理 ··· 14
 1.8　小结 ·· 17
 1.9　习题 ·· 17

第 2 章　流体的性质 ··· 19
 2.1　引言 ·· 19
 2.2　流体的基本性质 ··· 19
  2.2.1　流体压力 ·· 20
  2.2.2　温度 ·· 21
  2.2.3　密度 ·· 22
  2.2.4　黏性 ·· 23
  2.2.5　可压缩性 ·· 28
 2.3　流体的热力学性质 ··· 29
  2.3.1　比热 ·· 29
  2.3.2　比热比 ·· 30
  2.3.3　空气的导热性 ·· 31
 2.4　表面张力 ··· 32
 2.5　小结 ·· 34
 2.6　习题 ·· 34

第3章　流体静力学 ··········· 36

- 3.1　引言 ··········· 36
- 3.2　标量、矢量和张量 ··········· 36
- 3.3　体积力和表面力 ··········· 37
- 3.4　静止流体中的力 ··········· 38
- 3.5　流体微元上的压力合力 ··········· 39
- 3.6　流体静力学基本方程 ··········· 40
  - 3.6.1　流体静压分布 ··········· 41
  - 3.6.2　压力测量 ··········· 43
  - 3.6.3　压力测量单位和尺度 ··········· 46
- 3.7　大气 ··········· 46
  - 3.7.1　国际标准大气 ··········· 47
  - 3.7.2　平流层计算 ··········· 47
  - 3.7.3　对流层计算 ··········· 49
- 3.8　浸没表面上的静压力 ··········· 54
- 3.9　浮力 ··········· 56
- 3.10　小结 ··········· 57
- 3.11　习题 ··········· 58
- 参考文献 ··········· 62

第4章　流体运动学和动力学 ··········· 63

- 4.1　引言 ··········· 63
- 4.2　流体流动的描述 ··········· 63
  - 4.2.1　拉格朗日法和欧拉法 ··········· 63
  - 4.2.2　当地导数和随体（物质）导数 ··········· 64
  - 4.2.3　流体运动的图形化描述 ··········· 66
- 4.3　基本定律和补充定律 ··········· 68
  - 4.3.1　系统和控制体 ··········· 68
  - 4.3.2　积分分析和微分分析 ··········· 69
- 4.4　基本方程 ··········· 69
  - 4.4.1　连续性方程 ··········· 70
  - 4.4.2　动量方程 ··········· 70
  - 4.4.3　状态方程 ··········· 72
  - 4.4.4　边界层方程 ··········· 73
- 4.5　有旋和无旋运动 ··········· 75
  - 4.5.1　环量与涡量 ··········· 75
  - 4.5.2　流函数 ··········· 76
  - 4.5.3　流函数和速度势的关系 ··········· 77
- 4.6　位势流 ··········· 78

	4.6.1	二维的源和汇	81
	4.6.2	简单涡	82
	4.6.3	源-汇对	84
	4.6.4	偶极子	84
4.7	绕半体流动——简单流动的组合		88
4.8	小结		97
4.9	习题		97

第5章 流体动力学的几个问题 … 114

- 5.1 引言 … 114
- 5.2 黏性流动 … 114
- 5.3 物体受到的阻力 … 117
  - 5.3.1 压差阻力 … 118
  - 5.3.2 表面摩擦阻力 … 124
  - 5.3.3 不同物体阻力的比较 … 125
- 5.4 湍流 … 128
- 5.5 管道流动 … 136
- 5.6 无环量的圆柱绕流 … 142
- 5.7 有环量的圆柱绕流 … 146
- 5.8 小结 … 151
- 5.9 习题 … 151
- 参考文献 … 162

第6章 边界层 … 163

- 6.1 引言 … 163
- 6.2 边界层发展 … 164
- 6.3 边界层厚度 … 167
  - 6.3.1 位移厚度 … 168
  - 6.3.2 动量厚度 … 170
  - 6.3.3 动能厚度 … 171
  - 6.3.4 无量纲速度分布 … 172
  - 6.3.5 边界层类型 … 173
- 6.4 边界层流动 … 175
- 6.5 边界层求解 … 179
- 6.6 动量积分估算 … 179
  - 6.6.1 线性动量守恒 … 179
  - 6.6.2 平板边界层的卡门分析 … 181
- 6.7 无量纲形式边界层方程 … 182
- 6.8 平板边界层 … 189

　　　　6.8.1　层流边界层 ……………………………………………………………… 190
　　　　6.8.2　边界层厚度 …………………………………………………………… 192
　　6.9　沿平板不可压缩流动湍流边界层 …………………………………………… 201
　　6.10　有压力梯度的流动 …………………………………………………………… 205
　　6.11　层流积分理论 ………………………………………………………………… 206
　　6.12　小结 …………………………………………………………………………… 214
　　6.13　习题 …………………………………………………………………………… 214
　参考文献 ………………………………………………………………………………… 217

Chapter 1   Basic Concepts

第 1 章　基本概念

1.1　Introduction

Precise definitions of the basic concepts form the foundation for the proper development of a subject. Fluid mechanics has a unique vocabulary associated with it, like any other science. In this chapter, some important basic concepts associated with fluid mechanics are discussed. The unit systems and the law of dimensional homogeneity that will be used are also reviewed. Careful study of these concepts will be of great value for understanding the topics covered in the following chapters.

1.2　Some Basic Facts About Fluid Mechanics

Fluid mechanics may be defined as *the subject dealing with the investigation of the motion and equilibrium of fluids.* It is one of the oldest branches of physics and foundation for the understanding of many essential aspects of applied sciences and engineering. It is a subject of enormous interest in numerous fields such as biology, biomedicine, geophysics, meteorology, physical chemistry, plasma physics, and almost all branches of engineering.

Nearly two hundred years ago, man thought of laying down scientific rules to govern the motion of fluids. The rules were used mainly on the flow of water and air to understand them so that people can protect themselves from their fury during natural calamities such as cyclone and floods and utilize

1.1　引言

对基本概念的准确定义是一个学科正确发展的基础。和其他学科一样，流体力学也有一个与之相关的专业术语表。本章讨论与流体力学有关的一些重要基本概念，并对书中所涉及的单位制及量纲一致性原理进行综述。认真学习这些基本概念对后续章节内容的理解至关重要。

1.2　流体力学概况

流体力学可以被定义为一门**研究流体运动和平衡规律的学科**。流体力学是物理学古老的分支之一，也是理解应用科学和工程学许多基本原理的基础。在生物学、生物医学、地球物理学、气象学、物理化学、等离子体物理学等众多领域，以及几乎所有的工程学分支领域，流体力学都是非常受关注的学科。

大约两百年前，人们想到建立科学的定律来掌握流体的运动。这些定律主要用来理解水和空气的流动规律，从而使人类能够保护自己免受飓风和洪水等自然灾害的侵袭，并利用这些自然动力来发展土

their power to develop fields like civil engineering and naval architecture. In spite of the common origin, two distinct schools of thought gradually developed. On one hand, through the concept of "ideal fluid", mathematical physicists developed the theoretical science, known as *classical hydrodynamics*. On the other hand, realizing that idealized theories were of no practical application without empirical correction factors, engineers developed the applied science from experimental studies, known as *hydraulics*, for the specific fields of irrigation, water supply, river flow control, hydraulic power, and so on. Further, the development of aerospace, chemical, and mechanical engineering during the past few decades, and the exploration of space from 1960s have increased the interest in the study of fluid mechanics. Thus, it now ranks as one of the most-important basic subjects in engineering science.

The science of fluid mechanics has been extended into fields like regimes of hypervelocity flight and flow of electrical conducting fluids. This has introduced new fields of interest such as *hypersonic flow* and *magneto-fluid dynamics*. In this connection, it has become essential to combine the knowledge of thermodynamics, heat transfer, mass transfer, electromagnetic theory and fluid mechanics, for the complete understanding of the physical phenomenon involved in any flow process.

Fluid mechanics is one of the rapidly growing basic sciences, whose principles find application even in daily life. For instance, the flight of birds in air and the motion of fish in water are governed by the fluid mechanics rules. The design of various types of aircraft and ships is based on the fluid mechanics principles. Even natural phenomena like tornadoes and hurricanes can also be explained by the science of fluid mechanics. In fact, the science of fluid mechanics dealing with such natural

木工程和海事工程。尽管起源相同，但两个截然不同的流体力学学派逐渐发展了起来。一方面，数学物理学家们通过"理想流体"这一概念发展了流体力学理论科学，即**经典流体力学**。另一方面，工程师认识到不经过经验修正，理想化的理论就不实用，于是通过实验研究发展出用于灌溉、给水、河流控制、水力等专业领域的流体力学应用科学，即**水力学**。此外，航空航天、化学工程和机械工程在过去几十年的发展，以及20世纪60年代以来人类对外太空的探索，都进一步激发了人们对流体力学的研究热情。因此，流体力学已经成为工程科学中重要的基础学科之一。

流体力学已经拓展到超高速飞行和导电流体流动等范畴，并衍生出新的研究领域，例如**高超声速流动**和**磁流体动力学**。在这种拓展和演变背景下，将热力学、传热学、传质学、电磁理论与流体力学知识相结合，已经成为全面理解任何一个流动过程物理现象必不可少的手段。

流体力学是快速发展的基础科学之一，其原理甚至可应用于日常生活中。例如，空中鸟儿的飞翔和水中鱼儿的游动都符合流体力学规律。各种飞机和船舶的设计也要遵循流体力学原理。甚至像飓风和台风这样的自然现象也能用流体力学来解释。事实上，与这些自然现象相关的流体力学学科已经发展到能够很好地预测这些自然现象

phenomena has been developed to such an extent that they can be predicted well in advance. Since the earth is surrounded by an environment of air and water to a very large extent, almost everything that is happening on the earth and its atmosphere are some way or the other associated with the science of fluid mechanics.

The science of fluid motion is referred to as the mechanics of fluids, an allied subject of the mechanics of solids and engineering materials, and built on the same fundamental laws of motion. Therefore, unlike empirical hydraulics, it is based on the physical principles, and has close correlation with experimental studies which both compliment and substantiate the fundamental analysis, unlike the classical hydrodynamics which is based purely on mathematical treatment.

For understanding the fluid flows, it is essential to know the properties of fluids. Before we discuss the fluid properties, it will be useful if the difference between solids and fluids is understood. From the basic studies on physics, it is known that solids, liquids, and gases are the three states of matter. In general, liquids and gases are called fluids. It can be shown that this division into solid and fluid states constitutes a natural grouping of matter from the stand point of internal stresses and strains in elastic media, that is, the stress in a linear elastic solid is proportional to strain, while the stress in a fluid is proportional to its time rate of strain. In fact, among fluids themselves, only a group of fluids exhibits the above-said stress-strain relation and they are called Newtonian fluids. The above-mentioned behavior of solids and fluids may also be expressed in a simple way, as follows.

When a force is applied to a solid, deformation will be produced in the solid. If the force per unit area, viz. stress, is less than the yield

的程度。由于地球的绝大部分是被空气和水所包围的，因此地球上及大气中发生的一切几乎都或多或少和流体力学相关。

研究流体运动的科学被称为流体力学，是和固体及工程材料力学相并列的一个学科。这些学科建立在相同的基本运动定律基础上。因此，不像实验水力学，也不像建立在纯数学推导基础上的经典流体力学，流体力学建立在物理原理之上，并与用于支撑和验证基本分析的实验研究紧密结合。

为理解流体流动，首先应了解流体的特性。在讨论流体特性之前，先认清固体和流体之间的区别会很有帮助。由物理学基础研究可知，固体、液体和气体是物质的三种形态。通常，液体和气体统称为流体。这种按固体和流体形态进行的划分是从弹性介质内应力和应变的角度理所当然得到的物质分类，即线性弹性固体内部应力与应变成正比，而流体内应力与应变速率成正比。实际上，只有一部分流体表现出上述应力-应变关系，它们被称为牛顿流体。固体和流体的上述行为也可以简单描述如下。

当固体被施加作用力时，固体就会发生变形。如果单位面积上的作用力，即应力，小于材料的屈服

流体力学基础 Essence of Fluid Mechanics

stress, that is, within the proportional limit of the material, the deformation disappears when the applied force is removed. If the stress is more than the yield stress of the material, it will acquire a permanent setting or even break.

If a shearing force is applied to a fluid, it will deform continuously as long as the force is acting on it, regardless of the magnitude of the force.

The above difference in behavior between solids and liquids can be explained by their molecular properties. The existence of very strong intermolecular attractive forces in solids leads to their rigidity. The forces are comparatively weaker and very very small in liquids and gases, respectively. These characteristics enable liquid molecules to move freely within the liquid mass while still maintaining a close proximity to one another, whereas gas molecules have freedom to the extent of completely filling any space allotted to them.

The study of fluid flows can be classified into the following divisions.

Fluid statics, deals with fluid elements at rest with respect to one another and thus, free of shearing stress. The static pressure distribution in a fluid and on bodies immersed in a fluid can be determined from a static analysis.

Kinematics of fluid flow, deals with translation, rotation and rate of deformation of fluid particles. This analysis is useful in determining methods to describe the motion of fluid particles and analyzing flow patterns. However, the velocity and acceleration of fluid particles cannot be obtained from kinematic study alone, since the interaction of fluid particles with one another makes the fluid disturbed medium.

Fluid dynamics, deals with the determination of the effects of the fluid and its surroundings on the motion of the fluid. This involves the

应力，也就是在材料的比例极限范围内，去掉所施加的力后变形消失。如果应力大于材料的屈服应力，固体会产生永久变形，甚至断裂。

如果对流体施加剪切力，只要力是一直施加在流体上的，不管力的大小如何，流体都会一直发生变形。

上述固体和流体行为的区别可以用它们的分子特征来解释。固体中存在很强的分子间引力，使固体表现出刚性。而这一分子间作用力在液体和气体中分别表现为相对较弱和非常非常小。这些特性使液体分子能够在液体团内部自由移动，同时相互之间还能够保持紧密的距离；气体分子则能够自由地完全充满所在空间。

流体流动的研究可以分为如下几部分。

流体静力学：研究相互之间相对静止并因此不受剪切应力作用的流体微元。通过静态分析可以求得流体内部和浸没在流体内部的物体上的静态压力分布。

流体运动学：研究流体质点的移动、旋转和变形速率，用以确定流体质点运动的描述方法和分析流动形态。然而，流体质点间存在相互作用才使得流体能够被搅动起来，所以流体质点的速度和加速度不能仅从流体运动学分析中求得。

流体动力学：研究流体及其周围介质对流体运动的作用。流体动力学考虑相对运动流体质点上的作

consideration of forces acting on the fluid particles in motion with respect to one another. Since there is relative motion between fluid particles, shearing forces must be taken into consideration in the dynamic analysis.

1.3 Fluids and the Continuum

Fluid flows may be modeled at either microscopic or macroscopic level. The macroscopic model regards the fluid as a continuum, and the description is in terms of the variations of macroscopic velocity, density, pressure and temperature with distance and time. On the other hand, the microscopic or molecular model recognizes the particulate structure of a fluid as a myriad of discrete molecules and ideally provides information on the position and velocity of every molecule at all times.

The description of a fluid motion essentially involves a study of the behavior of all the discrete molecules which constitute the fluid. In liquids, the strong intermolecular cohesive forces make the fluid behave as a continuous mass of substance and, therefore, these forces need to be analyzed by the molecular theory. Under normal conditions of pressure and temperature, even gases have a large number of molecules in unit volume (for example, under normal conditions, for most gases, the molecular density is 2.7×10^{25} molecules per cubic meter) and, therefore, they also can be treated as a continuous mass of substance by considering the average effects of all the molecules within the gas. Such a fluid model is called continuum.

The continuum approach must be used only where it may yield reasonably correct results. For instance, this approach breaks down when the mean free path, the average distance traveled by the

用力。由于流体质点之间存在相对运动，因此流体动力学分析中必须考虑剪切力。

1.3 流体和连续介质

流体的流动可以在微观或宏观层面上建模。宏观模型把流体看作连续介质，以宏观的速度、密度、压力和温度随空间和时间的变化来描述流体的流动。而微观或分子模型把流体的具体结构看作无数个离散分子，并且理想化地给出所有时刻每个分子的位置和速度信息。

流体运动的描述本质上是对构成流体的所有离散分子整体运动行为的研究。在液体中，分子间较强的内聚力使得流体表现为一团连续的物质，因此应该利用分子理论对这些力进行分析。但在常温常压下，即使是气体，单位体积中也含有大量分子（例如，一般条件下，大多数气体中分子密度为 2.7×10^{25} 个/m³）。因此，通过考虑所有分子的平均效应，气体也可看成连续的物质。这样的流体模型被称为连续介质。

连续介质假设仅适用于能够得出合理结果的场合。例如，当分子的平均自由程——分子在两次连续碰撞之间的平均移动距离，和所研

molecules between two successive collisions, of the molecules is of the same order of magnitude as the smallest significant length in the problem being investigated. Under such circumstances, detection of meaningful, gross manifestation of molecules is not possible. The action of each molecule or group of molecules is then of significance and must be treated accordingly.

To understand this, it is essential to investigate the action of a gas on an elemental area inside a closed container. Even if the quantity of gas is assumed to be small, innumerable collisions of molecules on the surface result in the gross, non-time dependent manifestations of force. That is, the gas acts like a continuous substance. But if only a tiny amount of gas is kept in the container so that the mean free path is of the same order of magnitude as the sides of the area element, an erratic activity is experienced, as individual or groups of molecules bombard the surface. This cannot be treated as a constant force, and one must deal with an erratic force variation, as shown graphically in Figure 1.1.

究问题的最小特征长度在同一个数量级时，这个方法则不适用。在这种情况下，探究具有代表性的分子总体表现是不可能的，这时每个分子或分子群的作用更加重要，必须分别计算。

为理解这一点，研究某种气体对密闭容器内壁微元面积的作用是关键。即使气体的量很少，作用于容器内壁表面的无数次气体分子碰撞也会形成一个不随时间变化的合力。也就是说，气体也表现为连续介质的作用特点。但是，如果容器中只有微量的气体，使得平均自由程和微元面积的边长在同一数量级时，单个分子或分子群撞击壁面的行为就会变得不规则，这种行为不能被看作一个恒定的作用力，此时我们把这个力考虑成一个无规律的力的变化，如图1.1所示。

Figure 1.1　Force variation with time

图1.1　力随时间的变化

A continuous distribution of mass cannot exhibit this kind of variation. Thus, it is seen that in the first case the continuum approach would be applicable but in the second case the continuum approach would be questionable. From Figure 1.1,

一个连续分布的物质的作用力不会表现为这种变化（图1.1）的力的形式。可见在第一种情况中（容器中有少量气体），即使气体很少，连续介质方法也是适用的，但是在

it is clear that, when the mean free path is large in comparison to some characteristic length, the gas cannot be considered continuous and must be analyzed on the molecular scale. The mean free path of atmospheric air is between 50 nm and 70 nm. Another factor which influences the molecular activities of gas is the elapsed time between collisions. This time must be sufficiently small so that the random statistical nature of the molecular activity is preserved.

This book deals only with continuous fluids. Further, it will be assumed that the elastic properties are the same at all points in the fluid and are identical in all directions from any specified point. These stipulations make the fluid both *homogeneous* and *isotropic*.

1.4 The Perfect Gas Equation of State

Gases are basically divided into two broad categories, namely, *perfect* and *real gases*. A *perfect gas* is that in which intermolecular forces are negligible. A *real gas* is that in which intermolecular forces are important and must be accounted for.

In this book, we are concerned only with the fluids which can be regarded as perfect gases. For perfect gases, the kinetic theory of gases indicates that there exists a simple relation between pressure, specific volume, and absolute temperature. For a perfect gas at equilibrium, this relation has the following form:

$$pv = RT \tag{1.1}$$

This is known as *the perfect gas equation of state*. This equation is also called the ideal gas equation of state or simply the *ideal gas relation*, and a gas

第二种情况中只有极微量气体时，连续介质方法是有问题的。由图1.1可见，如果与某些特征长度相比平均自由程大，则气体不能被看作是连续的，而应在分子尺度上对它进行分析。大气的平均自由程在50nm到70nm范围内。另一个会影响气体分子运动的因素是碰撞之间所需的时间，这个时间必须足够小才能满足分子行为的随机统计规律。

本书仅针对连续介质的流体进行论述。尤其，本书假设变形特征在流体中各点和在某一点的各个方向上都是相同的，这些规定使得流体既具有**均一性**，又**各向同性**。

1.4 完全气体状态方程

气体基本上被分为**完全气体**和**实际气体**两大类型。**完全气体**是忽略分子间作用力的气体；**实际气体**是分子间作用力起重要作用从而必须加以考虑的气体。

本书中，我们只考虑可以被认为是完全气体的流体。对于完全气体，气体分子运动论指出气体的压力、比容和绝对温度之间存在一种简单的关系。对于平衡状态下的完全气体，这一关系有如下形式：

这就是**完全气体状态方程**，这一方程也称为理想气体状态方程或理想气体关系，符合这一关系的气体被

which obeys this relation is called an ideal gas. But it is essential to understand the difference between perfect and ideal gases. A perfect gas is calorically perfect. That is, its specific heats at constant pressure (c_p) and constant volume (c_v) are constants and independent of temperature. A perfect gas can be viscous or inviscid. Also, a perfect gas flow may be incompressible or compressible. But an ideal gas is assumed to be inviscid and incompressible. Therefore, we can state that, ideal gas is a special case of perfect gas. In Equation (1.1), p is the absolute pressure, T is the absolute temperature, v is the specific volume, and R is the gas constant. The gas constant R is different for each gas and is determined by	称为理想气体。但是必须理解完全气体和理想气体之间的区别，完全气体是量热完全气体，即它的定压比热（c_p）和定容比热（c_v）是常数，不随温度变化而变化。完全气体可以是有黏的或无黏的。同样，完全气体流可能是不可压缩的或可压缩的。而理想气体被假设为无黏且不可压缩。因此，理想气体可以看作一种特殊的完全气体。式（1.1）中，p 是绝对压力，T 是绝对温度，v 是比容，R 是气体常数。不同的气体，其气体常数 R 不同，可由下式计算：
$$R = R_u / M \quad [\text{kJ}/(\text{kg}\cdot\text{K})]$$	
or	或
$$R = R_u / M \quad [(\text{kPa}\cdot\text{m}^3)/(\text{kg}\cdot\text{K})]$$	
where R_u is the universal gas constant and M is the molar mass. The universal gas constant R_u is the same for all substances and its value is	式中，R_u 是通用气体常数；M 是摩尔质量。所有物质的通用气体常数都是相同的，它的值是
$$R_u = \begin{cases} 8.314\,\text{kJ}/(\text{kmol}\cdot\text{K}) \\ 8.314\,(\text{kPa}\cdot\text{m}^3)/(\text{kmol}\cdot\text{K}) \\ 0.08314\,(\text{bar}\cdot\text{m}^3)/(\text{kmol}\cdot\text{K}) \\ 1.987\,\text{Cal}/(\text{mol}\cdot\text{K}) \\ 8.31\,(\text{Pa}\cdot\text{m}^3)/(\text{kmol}\cdot\text{K}) \\ 62.3637\,(\text{mmHg}\cdot\text{m}^3)/(\text{kmol}\cdot\text{K}),\ 1\text{mmHg}=1.33322\times10^2\,\text{Pa} \end{cases}$$	
The molar mass M can be simply defined as the mass of one mole of a substance in grams, or the mass of one kmol in kilograms. It is essential to realize that an ideal gas is an imaginary substance that obeys the relation $pv = RT$. It has been experimentally observed that the ideal gas relation given above closely approximates the p-v-T behavior of real gases at low densities. At low pressures and high temperatures, the gas density decreases, and the gas behaves as an ideal gas.	摩尔质量 M 可简单定义为以克为单位的 1mol 物质的质量，或以千克为单位的 1000mol 物质的质量。 必须要认识到理想气体只是符合 $pv = RT$ 关系的假想物质。实验观测到上述理想气体关系近似于低密度实际气体的 p-v-T 特性。在低压和高温情况下，气体密度降低，气体则接近于理想气体状态。

Basically, state equation relates the pressure, density, and temperature of a substance. Although many substances are complex in behavior, experience shows that most gases of practical interest at moderate pressure and temperature are well represented by the perfect gas equation of state,

$$p = \rho RT \quad (1.2)$$

where ρ is the density of gas.

Although no real substance behaves exactly as an ideal or perfect gas, for air at room temperature, the perfect gas equation of state is in error by less than 1 percent for pressures up to 30atm(1atm=1.01325×10^5Pa). For air at 1atm, the equation is in error by less than 1 percent for temperatures as low as 140K.

In the range of practical interest, many familiar gases such as air, nitrogen, oxygen, hydrogen, helium, argon, neon, krypton and even heavier gases such as carbon dioxide can be treated as ideal gases with negligible error (often less than 1%). However, dense gases such as water vapor in steam power plants and refrigerant vapor in refrigerator should not be treated as ideal gases.

1.5 Regimes of Fluid Mechanics

Based on the flow properties which characterize the physical situation, the flows are classified into various types as follows.

1.5.1 Ideal Fluid Flow

It is only an imaginary situation where the fluid is assumed to be inviscid or non-viscous and incompressible. Therefore, there is no tangential force between adjacent fluid layers. An extensive mathematical theory is available for the ideal fluid.

状态方程初步建立了物质压力、密度和温度的关系。尽管很多物质的特性复杂，但经验表明大多数实际问题所涉及的气体在适当的压力和温度下都很符合完全气体状态方程：

式中，ρ 是气体密度。

尽管没有完全表现为理想气体或完全气体的实际物质，但对于室温下的空气，压力高达 30atm（1atm=1.01325×10^5Pa）时，完全气体状态方程的误差小于 1%；对于 1atm 下的空气，温度低至 140K 时，方程的误差也小于 1%。

在实际研究范畴，许多熟悉的气体，例如空气、氮气、氧气、氢气、氦气、氩气、氖气、氪气，甚至更重的气体，如二氧化碳，在可忽略的误差范围内（通常小于 1%）都可以看成理想气体。然而，密度大的气体如蒸汽发电站中的水蒸气和冰箱中的制冷蒸气就不能看作理想气体。

1.5 流体力学范畴

根据流动特性所体现的物理状态，流动可以被分为如下几种类型。

1.5.1 理想流体流动

理想流体流动只是一种把流体假设为非黏的或无黏的而且不可压缩的流体而假想的流动状态。因此，相邻的流体层之间没有切向力。大量的数学理论都适用于理想

Although the theory of ideal fluids fails to account for viscous and compressibility effects in actual fluid flow processes, it gives reasonably reliable results in the calculation of lift, induced drag and wave motion for gas flow at low velocity and for water. This branch of fluid mechanics is called Classical Hydrodynamics.

1.5.2 Viscous Incompressible Flow

The theory of viscous incompressible fluids assumes fluid density to be constant. It finds widespread application in the flow of liquids and the flow of air at low velocity. The phenomena involving viscous forces, flow separation and eddy flows are studied with the help of this theory.

1.5.3 Gas Dynamics

Gas dynamics may be defined as the science of flow field where a change in pressure is accompanied by density and temperature changes. The theory of gas dynamics deals with the dynamics and thermodynamics of the flow of compressible fluids. Based on the dimensionless velocity, namely Mach number, *Ma*, defined as *the ratio of flow velocity and the local speed of sound*, gas dynamics can be further divided into the fields of study commonly referred to as subsonic ($Ma<1$), *transonic* ($Ma \approx 1$), *supersonic* ($1<Ma \leqslant 5$), and *hypersonic* ($Ma>5$) gas dynamics. The classification between supersonic and hypersonic flow is arbitrary because it depends on the shape of the body over which the fluid flows. In a gas dynamic flow a change in velocity is accompanied by a change of density and temperature. Thus, gas dynamics is essentially a combination of fluid mechanics and thermodynamics. It is referred to as *aerothermodynamics*, when it deals with the aerodynamic forces and moments and heating distribution of a vehicle that flies at

流体，尽管理想流体理论不能用于计算实际流体流动过程中黏性和可压缩性的影响，但对于低速气体流动和水中升力、阻力以及波运动的计算，这一理论也能够给出合理可靠的结果。这一流体力学分支称为经典流体力学。

1.5.2 黏性不可压缩流动

黏性不可压缩流体理论假设流体密度恒定，这一理论广泛应用于液体流动和低速气体流动中，涉及黏滞力、流动分离和涡流的现象都可以用这一理论进行研究。

1.5.3 气体动力学

气体动力学可以定义为研究流场中压力变化时，伴随密度和温度变化的流场的科学。气体动力学理论涉及可压缩流体流动的动力学和热力学，在马赫数 Ma 被定义为**流动速度和当地声速的比值**这一无量纲速度基础上，气体动力学通常可以进一步分为**亚声速**（$Ma<1$）、**跨声速**（$Ma \approx 1$）、**超声速**（$1<Ma \leqslant 5$）和**高超声速**（$Ma>5$）气体动力学研究领域。超声速和高超声速之间的划分还有其他的数值，因为这一划分还与流体所流经物体的形状有关。在气体流动中，速度变化伴随着密度和温度变化。因此，气体动力学本质上是流体力学和热力学的结合。当求解高超声速飞行器的气动力、动量热分布时，也称为**气动热力学**。对于气体动力学，除了马赫数外，普朗特数和比热比 $\gamma (=c_p/c_v)$ 也是很重要的控

hypersonic speeds. For gas dynamics, in addition to Mach number, Prandtl number and the ratio of specific heats γ (= c_p/c_v) also play an important role, being control parameters. The Prandtl number, Pr, which is the ratio of kinematic viscosity and thermal diffusivity, is a measure of the relative importance of velocity and heat conduction.

Regardless of speed ranges, the theory of gas dynamics can be divided into two parts, inviscid gas dynamics and viscous gas dynamics. The inviscid theory is important in the calculation of nozzle characteristics, shock waves, lift and wave drag of a body, while the viscous theory is applicable to the calculation of skin friction and heat transfer characteristics of a body moving through a gas, such as atmospheric air.

1.5.4 Rarefied Gas Dynamics

The concept of continuum fails when the mean free path of fluid molecules is comparable to some characteristic geometrical parameters in the flow field. A dimensionless parameter, Knudsen number, Kn, defined as the ratio of mean free path to a characteristic length, aptly describes the degree of departure from continuum flow. Based on the Knudsen number, the flow regimes are grouped as

Continuum Flow (Kn<0.01): All equations of viscous compressible flow are applicable in this regime. The no-slip boundary condition is valid.

Slip Flow (0.01≤Kn≤0.1): Here again the continuum fluid dynamic analysis is applicable provided the slip boundary conditions are employed. That is, the no-slip boundary condition of continuum flows, dictating zero velocity at the surface of an object kept in the flow, is not valid. The fluid molecules move (slip) with a finite velocity, called slip velocity, at the boundary.

Transition Flow (0.1<Kn≤10): In this regime

制参数。普朗特数 Pr 作为动力黏度和热耗散率之比，是对速度和热传导相对重要程度的衡量。

如果不考虑速度范围，气体动力学理论可被分为两部分：无黏气体动力学和黏性气体动力学。无黏理论主要在喷嘴特性、激波、物体升力和波阻力的计算中起重要作用，而黏性理论主要用于计算物体在气体例如大气中运动时所受到的表面摩擦力和传热特性。

1.5.4 稀薄气体动力学

当流体分子的平均自由程与流场中一些特征几何参数差不多时，连续介质概念不成立。一个无量纲参数，被定义为平均自由程和特征长度之比的克努森数 Kn，能够描述流体与连续介质流动不符的程度。基于克努森数，流动范畴可分为如下几组。

连续介质流（Kn<0.01）：所有黏性可压缩流动的方程都适用于这一流动范畴，无滑移边界条件成立。

滑流（0.01≤Kn≤0.1）：对于这一流动，如果采用滑移边界条件，连续流体动力学分析仍然适用。即假定物体表面的速度在流动中恒定为零的连续介质无滑移边界条件是不适用的。流体分子在边界处以被称为滑移速度的有限速度移动。

过渡流（0.1<Kn≤10）：在这

of flow, the fluid cannot be treated as continuum. At the same time, it cannot be treated as a free molecular flow since such a flow demands the intermolecular force of attraction to be negligible. Hence, it is a flow regime between continuum and free molecular. The kinetic theory of gases must be employed to adequately describe this flow.

Free Molecular Flow ($Kn > 10$): In this regime of flow the fluid molecules are so widely dispersed that the intermolecular forces cannot be neglected.

All these regimes of rarefied gas dynamics or super aerodynamics are encountered at high altitudes, where the molecular density is very low. This branch of fluid flow is also called low-density flow.

Magnetofluid mechanics: The subject of magnetofluid mechanics is an extension of fluid mechanics, with thermodynamics, mechanics, materials and the electrical sciences. This branch was initiated by astrophysicists. Other names which are used to refer to this discipline are magnetohydrodynamics, magnetogasdynamics and hydromagnetics.

Magnetofluid mechanics is the study of the motion of an electrically charged conducting fluid in the presence of a magnetic field. The motion of an electricity conducting fluid in the magnetic field will induce electric currents in the fluid, thereby modifying the field. The flow field will also be modified by the mechanical forces produced by it. The interaction between the field and the motion makes magneto fluid dynamics analysis difficult.

A gas at normal and moderately high temperature is a non-conductor. But at very high temperatures, of the order of 10000K and above, thermal excitation sets in. This leads to dissociation and ionization. Ionized gas is called plasma, which is an electrically conducting medium. Electrically con-

种流动中，流体不能看作是连续的，同时，因为这种流动要求忽略分子间的吸引力，也不能看成自由分子流动，因此是介于连续和自由分子之间的流动，必须采用气体分子运动论方法才足以描述这种流动。

自由分子流（$Kn>10$）：这一流动范畴中流体分子非常分散以致分子间作用力不能被忽略。

在分子密度极低的高海拔地区会遇到所有这类稀薄气体动力学或超高空空气动力学的情况。这一流体流动分支也被称为低密度流动。

磁流体力学：磁流体力学学科是流体力学与热力学、力学、材料学以及电子学相结合的一个延伸学科，这个分支由航天物理学家始创，也称为磁液动力学、磁气动力学和磁流体动力学。

磁流体力学研究被充电的导电流体在磁场中的运动规律，导电流体在磁场中的运动会在流体中产生电流，因此会改变磁场。流场也会被磁场产生的机械力所改变，磁场和流动的相互作用使得磁流体力学分析十分困难。

常温或稍高温度下的气体是不导电体，但是在高达 10000K 数量级的高温或更高温度下，就会开始热激发，从而导致解离或电离。被离子化的气体称为等离子体，是一种导电介质。在导弹和飞船重返

ducting fluids are encountered in engineering problems like re-entry of missiles and spacecraft, plasma jet, controlled fusion research and magneto-hydrodynamic generator.

1.5.5 Flow of Multicomponent Mixtures

This field is simply an extension of basic fluid mechanics. The analysis of flow of homogeneous fluid consisting of single species, termed basic fluid mechanics, is extended to study the flow of chemically reacting component mixtures, made of more than one species. All the three transports, namely, the momentum transport, energy transport, and mass transport, are considered in this study, unlike the basic fluid mechanics where only transport of momentum and energy are considered.

1.5.6 Non-Newtonian Fluid Flow

Fluids for which the stress is not proportional to time rate of strain are called *non-Newtonian fluids*. Such fluids show a nonlinear dependence of shearing stress on velocity gradient. Examples of non-Newtonian fluids are, honey, printers ink, paste and tar.

1.6 Dimension and Units

In fluid dynamics, mostly the gross, measurable molecular manifestations such as pressure and density as well as other equally important, measurable abstract entities, for example, length and time, will be dealt with. These manifestations which are characteristics of the behavior of a particular fluid, and not of the manner of flow, may be called *fluid properties*. Density and viscosity are example of fluid properties. In order to adequately discuss these properties, a consistent set of standard units must be

大气、等离子喷射、受控熔接研究和磁液动力发生器等工程问题中都会遇到导电流体。

1.5.5 多元混合流动

这一领域只是基础流体力学的一个延伸，单一流体（包含单一介质）流动的分析称为基础流体力学，可延伸到研究化学反应多元混合物（由不止一种介质组成）的流动。与只考虑动量或能量传递的基础流体力学不同，这种多元混合流动的研究要同时考虑三种传输形式，即动量传输、能量传输和质量传输。

1.5.6 非牛顿流体流动

应力与应变率不成正比的流体称为**非牛顿流体**。这种流体表现为剪切应力随速度梯度呈非线性变化，例如蜂蜜、打印机墨水、胶和焦油等都是非牛顿流体。

1.6 量纲和单位制

流体动力学中，通常使用体现分子运动的表征参数，如压力和密度以及另一些同样重要的可测量的特征参数如长度和时间。这些表征量是表示某一具体流体行为的特征，而不是流动方式的表现，它们可以称为**流体性质**。例如密度和黏度就是流体性质。为了充分讨论这些性质，必须定义一套统一的标准单位制。表1.1给出了常用

defined. Table 1.1 gives the common system of units and their symbol.

单位制。

Table 1.1 Common systems of units
表 1.1 常用单位制

Quantity 物理量	SI 国际单位制	CGS 厘米-克-秒单位制	FPS 英尺-磅-秒单位制	MKS 米-千克-秒单位制
mass 质量	kg	g	lb	kg
length 长度	m	cm	ft	m
time 时间	s	s	s	s
force 力	N	dyn	pdl	kgf
temperature 温度	K	℃	°F	℃

In this text, throughout we shall use the SI system of units. However, other systems of units mentioned above are equally applicable to all the equations.

本书全部采用 SI，但上述其他单位制也同样适用于所有方程。

1.7 Law of Dimensional Homogeneity

1.7 量纲一致性原理

This law states, "an analytically derived equation representing a physical phenomenon must be valid for all system of units". Thus, the equation for the frequency of a simple pendulum, $f = (1/2\pi)\sqrt{g/l}$, is properly stated for any system of units. This explains why all natural phenomena proceed completely in accordance with man-made units, and hence fundamental equations representing such events should have validity for any system of units. Thus, the fundamental equations of physics are dimensionally homogeneous, and consequently all relations derived from these equations must also be dimensionally homogeneous. For this to occur under all systems of units, it is necessary that each

这个原理指出"凡是根据基本物理规律导出的物理方程必须适用于所有单位制"。因此，表示单摆频率的方程 $f = (1/2\pi)\sqrt{g/l}$ 就是适用于所有单位制的描述。这一原理就解释了为什么所有自然现象的发展会完全符合人为制定的单位制，而且描述这些现象的基本方程应该适用于所有单位制。所以，物理学基本方程是量纲一致的，继而所有从这些方程推导出的关系式也都应该具有量纲一致性。为此，方程中的各项都必须具有同样的量纲表达式。

grouping in an equation has the same dimensional representation.

Examine the following dimensional representation of an equation:

$$L = t^2 + t$$

where L denotes length and t is the time. Changing the units of length from feet to meters will change the value of the left-hand side while not affecting that of the right-hand side, thus making the equation invalid in the new system of units. Dimensionally homogeneous equations only will be considered in this book.

In 1872 an international meeting in France proposed a treaty called the Metric Convention, which was signed in 1875 by 17 countries including the United States. It was an important step over British systems because its use of base 10 is the foundation of our number system. Problems still remained because even the metric countries differed in their use of kilo-pounds instead of dynes or newton, kilograms instead of grams, or calories instead of joule. To standardize the metric system, a General Conference of Weights and Measures attended by 40 countries in 1960 proposed the International Systems of Units (SI units). The SI units become more and more popular throughout the world and it is expected to replace all other systems of units in due course.

The International System of Units (abbreviated SI from French: Le Système International D'unités) is the modern form of the metric system and is the world's most widely used system of measurement, used in both everyday commerce and science. It comprises a coherent system of units of measurement built around 7 base units, 22 named and an indeterminate number of unnamed coherent derived units, and a set of prefixes that act as

观察一个关系式的如下量纲表达式：

式中，L 是长度；t 是时间。如果把长度单位从英尺改为米会改变等式左边的数值，但不会影响右边数值，这样就会使得关系式在新单位制中不成立。本书只采用具有量纲一致性的关系式。

1872 年，在法国举办的一个国际会议提出了一个称为米制公约的议案，1875 年这一议案由包括美国在内的 17 个国家共同签署。因为这一公约所采用的十进制是数字系统的基础，因此这成为跨越英制的重要进步。但即使是采用米制的国家在单位制的使用方面也存在差别，例如在使用千磅而不是达因或牛顿、千克而不是克、卡路里而不是焦耳方面的问题仍然存在。为了使米制标准化，1960 年有 40 个国家参加的国际计量大会提出了国际单位制（SI），于是国际单位制在全世界范围越来越普遍，并且有望在将来取代所有的其他单位制。

国际单位制（SI 来自于法语的 Le Système International D'unités）是米制的最新形式，也是全世界日常贸易和科学中最广泛使用的度量制。这一单位制是一套建立在大约 7 个基本单位、22 个已命名的和许多未命名的相关导出单位以及一套用作十进倍数的词头基础上的计量单位制，是国际量制的组成

decimal-based multipliers. It is one part of the International System of Quantities.

As we see, there are four primary dimensions in fluid mechanics from which all other dimensions can be derived. They are mass, length, time, and temperature. These variables and their units are given in Table 1.1. All other variables in fluid mechanics can be expressed in terms of dimensions M, L, T and Θ for the variables of mass, length, time and temperature, respectively. For instance, velocity has the dimensions LT^{-1}. The dimension for any variable Q can be expressed as $\dim Q$. One of the interesting derived dimensions is that of force, which is directly related to mass, length, and time. By Newton's second law,

如前所述，流体力学中有四个基本量纲——质量、长度、时间和温度，其他量纲都可由它们推导而来。这些变量和它们的单位如表1.1所示。流体力学中所有量都可以用质量、长度、时间和温度的量纲 M、L、T 和 Θ 表示出来，例如速度的量纲 LT^{-1}。任意变量 Q 的量纲可表示为 $\dim Q$。一个有意思的导出量纲就是力，可直接由质量、长度以及时间建立联系。牛顿第二定律为

$$F = ma \tag{1.3}$$

where m is mass and a is acceleration. From this relation it is seen that, dimensionally, $\dim F = MLT^{-2}$. A constant of proportionality which will figure in the above force relation is avoided by defining the force unit exactly in terms of the primary units. Thus, the newton can be defined as

式中，m 是质量；a 是加速度。从这一关系可见，量纲上，有 $\dim F = MLT^{-2}$。如果完全以基本单位制的形式定义力的单位，就不需要在上述力关系中使用比例常数了。因此，牛顿这个单位可定义为

$$1 \text{ newton of force}（1牛顿的力）= 1\text{N} = 1(\text{kg} \cdot \text{m})/\text{s}^2 \tag{1.4}$$

Some of the derived variables which are often used in the study of the fluid dynamics along with their dimensions in terms of the primary dimensions and units are given in Table 1.2.

对于流体动力学研究中一些常用的导出量，用基本量纲形式表示的这些量的量纲及单位如表1.2所示。

Table 1.2 Units and dimensions of some derived variables
表 1.2 一些导出量的量纲和单位

Variables 量	Dimensions 量纲	SI Unit SI 单位
Acceleration 加速度	LT^{-2}	m/s²
Angular Velocity 角速度	T^{-1}	s⁻¹
Area 面积	L^2	m²

续表

Variables 量	Dimensions 量纲	SI Unit SI 单位
Density 密度	ML^{-3}	kg/m^3
Energy, Heat, Work 能量，热量，功	ML^2T^{-2}	$J=N\cdot m$
Power 功率	ML^2T^{-3}	$W=J/s$
Pressure or Stress 压力或应力	$ML^{-1}T^{-2}$	$Pa=N/m^2$
Velocity 速度	LT^{-1}	m/s
Viscosity 黏度	$ML^{-1}T^{-1}$	$kg/(m\cdot s)$
Volume 体积	L^3	m^3

1.8 Summary

This chapter introduces some basic facts about fluid mechanics. Fundamental concepts for fluid mechanics, especially the difference between solids and fluids, as well as perfect and real gases, are provided. The regimes of fluid mechanics are talked about. The Law of Dimensional Homogeneity is given to help the understanding of equations and mathematical descriptions for the future study.

1.9 Exercises

Problem 1.1 Determine the gas constants of (a) hydrogen, (b) oxygen and (c) nitrogen.

[Ans: (a) $4136.32 m^2/(s^2\cdot K)$,
(b) $259.81 m^2/(s^2\cdot K)$,
(c) $296.93 m^2/(s^2\cdot K)$]

Problem 1.2 Determine the density of air at standard sea level conditions.

[Ans: $1.225 kg/m^3$]

1.8 小结

本章介绍了流体力学概况，引入了流体力学的基本概念，尤其是固体和流体，以及完全气体和实际气体的区别。讨论了流体力学的研究范畴，并给出了量纲一致性原理，以帮助读者更好地理解后续学习中的公式和数学描述。

1.9 习题

题 1.1 求（a）氢气、（b）氧气和（c）氮气的气体常数。

【答：（a） $4136.32 m^2/(s^2\cdot K)$，
（b） $259.81 m^2/(s^2\cdot K)$，
（c） $296.93 m^2/(s^2\cdot K)$】

题 1.2 求标准海平面条件下空气的密度。

【答： $1.225 kg/m^3$】

Problem 1.3 During a study of a certain flow system the following equation relating the pressure p_1 and p_2 at two points is developed.

$$p_2 = p_1 + \frac{flV}{Dg}$$

where V is a velocity, l is the distance between two points, D is a diameter, g is the gravitational acceleration, and f is a dimensionless coefficient. Is the equation dimensionally consistent?

[Ans: No]

题 1.3 某一流动系统研究中,两点之间压力 p_1 和 p_2 可建立如下关系式:

式中,V 是速度;l 是两点之间的距离;D 是直径;g 是重力加速度;f 是无量纲系数。该关系式是否量纲一致?

【答:否】

Chapter 2　Properties of Fluids

第 2 章　流体的性质

2.1　Introduction

Gases and liquids are generally termed fluids. Though the physical properties of gases and liquids are different, they are grouped under the same heading since both can be made to flow unlike the solid. This chapter mainly introduces the basic properties of fluids, including the pressure, temperature, density and viscosity. The thermodynamic properties and surface tension are also described.

Under dynamic conditions, the nature of governing equations is the same for both gases and liquids. Hence, it is possible to treat them under the same heading, namely, fluid dynamics or fluid mechanics. However, certain substances known as viscoelastic materials behave like liquid as well as solid, depending on the rate of application of the force. Pitch and silicone putty are typical examples of viscoelastic material. If the force is applied suddenly the viscoelastic material will behave like a solid, but with gradually applied pressure the material will flow like the liquid. The properties of such materials are not considered in this book. Similarly, non-Newtonian fluids, low-density flows, and two-phase flows such as gas liquid mixtures are also not considered in this book. The experimental techniques described in this book are for the well-behaved simple fluids such as air.

2.2　Basic Properties of Fluids

Fluid may be defined as a substance which will

2.1　引言

气体和液体统称为流体。尽管二者物理性质不同，但不同于固体，气体和液体由于都可流动，因此可归在同一章中进行讨论。本章主要介绍流体的各种基本性质，包括压力、温度、密度和黏度，以及流体的热力学特性和表面张力。

在运动条件下，气体和液体控制方程的本质是相同的，因此，把二者放在同一标题下，即流体动力学或流体力学下进行讨论是可行的。然而，某些被称为黏弹性材料的物质，随施加力的速率不同，可表现为既像液体又像固体，如沥青和硅油就是黏弹性材料的典型实例。如果突然施加作用力，黏弹性材料将表现得像固体，然而逐渐施加压力时，黏弹性材料则会像液体一样流动。这类黏弹性材料的性质不在本书的介绍范围。同样，非牛顿流体、低密度流体和气液混合这样的两相流动也都不在本书中介绍。本书讲述的实验技术也只针对例如空气这样特性明显的简单流体。

2.2　流体的基本性质

流体可以被定义为只要受到

continue to change shape as long as there is a shear stress present, however small it may be. That is, the basic feature of a fluid is that it can flow, and this is the essence of any definition of it. Examine the effect of shear stress on a solid element and a fluid element, shown in Figure 2.1.

It is seen from this figure that the change in shape of the solid element is characterized by an angle $\Delta\alpha$, when subjected to a shear stress. Whereas, for the fluid element there is no such fixed $\Delta\alpha$, even for an infinitesimal shear stress. A continuous deformation persists as long as shearing stress is applied. The rate of deformation, however, is finite and is determined by the applied shear force and the fluid properties.

剪切应力作用，无论所受剪切应力多小，都会连续地改变形状的一种物质，即流体的基本特征是能够流动，这也是它所有定义的本质。观察一个固体微元和一个流体微元上剪切应力的作用结果，如图 2.1 所示。

该图表明固体微元的形变是用剪切应力作用下产生的一个角度 $\Delta\alpha$ 来描述的。然而，对于流体微元，即使是施加一个无限小的剪切应力，也不存在这样一个固定的角度 $\Delta\alpha$。只要对流体施加剪切应力，变形就会一直持续下去。但是，变形速率是有限的，其取决于所施加的剪切力和流体的性质。

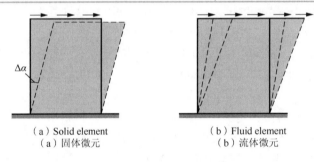

(a) Solid element　　　　　(b) Fluid element
(a) 固体微元　　　　　　(b) 流体微元

Figure 2.1　Solid and fluid elements under shear stress

图 2.1　剪切应力作用下的固体和流体微元

2.2.1　Pressure of Fluids

Pressure may be defined as the force per unit area which acts normal to the surface of any object which is immersed in a fluid. For a fluid at rest, at any point the pressure is the same in all directions. The pressure in a stationary fluid varies only in the vertical direction, and is constant in any horizontal plane. That is, in stationary fluids the pressure increases linearly with depth. This linear pressure distribution is called hydrostatic pressure distribution.

2.2.1　流体压力

流体压力可定义为流体垂直作用于浸没在流体中任一物体单位表面积上的力。对于静止流体，在任意点处各个方向上的压力大小相等。静止流体中，压力仅在垂直方向上变化，在任意水平面内，压力都是常值，即静止流体中，压力随深度线性增加。这一线性压力分布称为流体静压分

The hydrostatic pressure distribution is valid for moving fluids, provided there is no acceleration in the vertical direction. This distribution finds extensive application in manometry.

When a fluid is in motion, the actual pressure exerted by the fluid in the direction normal to the flow is known as the static pressure. If there is an infinitely thin pressure transducer which can be placed in a flow field without disturbing the flow, and made to travel with the same speed as that of the flow then it will record the exact static pressure of the flow. From this stringent requirement of the probe for static pressure measurement, it can be inferred that exact measurement of static pressure is impossible. However, there are certain phenomena, such as "the static pressure at the edge of a boundary layer is impressed through the layer", which are made use of for the proper measurement of static pressure. The pressure which a fluid flow will experience if it is brought to rest, isentropically, is termed total pressure. The total pressure is also called impact pressure. The total and static pressures are used for computing flow velocity.

Since pressure is intensity of force, it has the dimensions

布。如果在垂直方向上没有加速度，静压力分布规律也适用于移动的流体，因此这一分布规律广泛应用于压力的测量中。

当流体流动时，其产生的垂直于流动方向的实际压力称为静压。如果存在一个能放置在流场中而不干扰流动的无限薄测压探头，和流体一起以相同的速度运动，则该测压探头记录的就是流动的静压。由对静压测量探头的这一严苛要求，可想而知静压的精确测量几乎是不可能的。然而，存在某些现象，例如"边界层内部静压与边界层边缘处静压保持一致"，可以被用来正确地测量静压。如果流体流动被等熵地止动，流体将要承受的压力称为总压，总压也叫冲击压。总压和静压的概念可被用来计算流速。

压力是力的强度，因此有如下量纲：

$$\dim \frac{F}{A} = \frac{\mathrm{MLT}^{-2}}{\mathrm{L}^2} = \mathrm{ML}^{-1}\mathrm{T}^{-2}$$

and is expressed in the units of Newton per square meter (N/m^2) or simply pascal (Pa). At standard sea level condition, the atmospheric pressure is 101325Pa, which corresponds to 760mmHg.

并且压力单位用牛顿每平方米（N/m^2）来表示，或简单地表示为帕斯卡（Pa）。在标准海平面条件下，大气压为101325Pa，对应为760mmHg。

2.2.2 Temperature

In any form of matter, the molecules are continuously moving relative to each other. In gases the molecular motion is a random movement of appreciable amplitude ranging from about 76×10^{-9}m, under normal conditions (that is, at standard sea level

2.2.2 温度

在任何形态的物质中，分子都相对于彼此连续不断地运动着。在气体中，分子运动是振幅变化范围很大的随机运动，振幅约从正常情况下（即在标准海平

pressure and temperature), to some tens of millimeters, at very low pressures. The distance of free movement of a molecule of a gas is the distance it can travel before colliding with another molecule or the walls of the container. The mean value of this distance for all molecules in a gas is called the molecular mean free path length. By virtue of this motion the molecules possess kinetic energy, and this energy is sensed as temperature of the solid, liquid or gas. In the case of a gas in motion, it is called the static temperature. Temperature has units Kelvin (K) or degree celsius (℃), in SI units. For all calculations in this book, temperature will be expressed in Kelvin, that is, from absolute zero. At standard sea level condition, the atmospheric temperature is 288.15K.

2.2.3 Density

The total number of molecules in a unit volume is a measure of the density, ρ, of a substance. It is expressed as mass per unit volume, say kg/m^3. Mass is weight divided by acceleration due to gravity. At standard atmospheric temperature and pressure (288.15K and 101325Pa, respectively), the density of dry air is 1.225kg/m^3.

Density of a material is a measure of the amount of material contained in a given volume. In a fluid system, the density may vary from point to point. Consider the fluid contained within a small spherical region of volume δV, centered at some point in the fluid, and let the mass of fluid within this spherical region be δm. Then the density of the fluid at the point on which the sphere is centered can be defined by

$$\rho = \lim_{\Delta V \to 0} \frac{\delta m}{\delta V} \qquad (2.1)$$

There are practical difficulties in applying the above definition of density to real fluids composed of discrete molecules, since under the limiting condition the sphere may or may not contain any molecule. If it contains, say, just a single molecule, the value obtained for the density will be fictitiously high. If it does not contain any molecule the resultant value of density will be zero. This difficulty can be avoided over the range of temperatures and pressures normally encountered in practice, by the following two ways.

(1) The molecular nature of a gas may be ignored, and the gas is treated as a continuous medium or continuous expanse of matter, termed continuum (that is, does not consist of discrete particles).

(2) The decrease in size of the imaginary sphere may be assumed to reach a limiting size, such that, although it is small compared to the dimensions of any physical object present in a flow field, for example an aircraft, it is large enough compared to the fluid molecules and, therefore, contains a reasonably large number of molecules.

2.2.4 Viscosity

The property which characterizes the resistance that a fluid offers to applied shear force is termed viscosity. This resistance, unlike for solids, does not depend upon the deformation itself but on the rate of deformation. Viscosity is often regarded as the stickiness of a fluid and its tendency is to resist sliding between layers. There is a very little resistance to the movement of the knife-blade edge-on through air, but to produce the same motion through a thick oil needs much more effort. This is because the viscosity of the oil is higher compared to that of air.

上述密度定义应用于由离散分子组成的真实流体时会存在一些实际的困难。因为在极限情况下，这个球可能包含也可能不包含任何分子。如果它仅包含一个单分子，则求得的密度值将比想象中高很多。如果它不包含任何分子，密度计算值将是零。对于实际应用中通常可能会遇到的温度和压力范围，这个困难可以通过以下两种方式来避免。

（1）可以忽略气体的分子本质，将气体视为一种连续的介质，或连续延伸的物质，称为连续介质（即不含有离散的粒子）。

（2）可以把这个假想的球减小到一个有限的尺寸，例如，尽管相对于流场中任何实际物体的尺寸，比如一架飞机的尺寸而言，这个尺寸是很小的，但相比于流体分子，它又是足够大的，这样，这个尺寸就能够包含足够多数量的分子。

2.2.4 黏性

当受到剪切力作用时流体会产生抵抗这一剪切力作用的阻力，这种性质称为黏性。不同于固体，流体的这种阻力与变形本身无关，而是取决于变形速率。黏性通常被看作流体的黏滞性，其趋势是阻止流体层之间的相对滑动。刀刃在空中划过时受到很小的阻力，但是在黏稠的油里，产生同样的运动就需要更大的力，这是因为油的黏性比空气黏性大。

1. Coefficient of Absolute Viscosity

The coefficient of absolute viscosity is a direct measure of the viscosity of a fluid, which is also called "dynamics viscosity". Consider the two parallel plates placed at a distance h apart, as shown in Figure 2.2(a).

1. 绝对黏度

绝对黏度是流体黏性的一种直接度量，也称为"动力黏度"。考察两块间距为 h 的平行平板，如图2.2（a）所示。

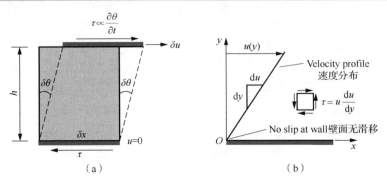

Figure 2.2　Fluid shear between a stationary and a moving parallel plate

图2.2　静止和移动平行平板之间的流体剪切

The space between them is filled with a fluid. The bottom plate is fixed and the other is moved in its own plane at a speed u. The fluid in contact with the lower plate will be at rest, while that in contact with the upper plate will be moving with speed u, because of no-slip condition. In the absence of any other influence, the speed of the fluid between the plates will vary linearly, as shown in Figure 2.2(b). As a direct result of viscosity, a force F has to be applied to each plate to maintain the motion, since the fluid will tend to retard the motion of the moving plate and will tend to drag the fixed plate in the direction of the moving plate. If the area of each plate in contact with fluid is A, then the shear stress acting on each plate is F/A. The rate of sliding of the upper plate over the lower is u/h.

These quantities are connected by Maxwell's

两平板之间的空间充满了流体。底部平板固定，另一块平板在自己的平面内以速度 u 移动。根据无滑移条件，与下平板接触的流体会静止，而与上平板接触的流体将以速度 u 移动。如果没有任何其他影响因素，平板之间流速将线性变化，如图2.2（b）所示。由于黏性的作用，流体要阻止运动平板的移动，并向运动平板的移动方向拖动固定平板，因此必须要施加一个力 F 以维持两平板的各自运动。如果接触流体的平板面积是 A，则作用在每个平板上的剪切应力为 F/A。上平板相对于下平板的滑移率为 u/h。

麦克斯韦方程把上述各量联

equation, which serves to define the coefficient of absolute viscosity μ. Maxwell's definition of viscosity states that, "the coefficient of viscosity is the tangential force per unit area on either of two parallel plates at unit distance apart, one fixed and the other moving with unit velocity".

Maxwell's equation for viscosity is

$$\frac{F}{A} = \mu \frac{u}{h} \qquad (2.2)$$

Hence,

$$ML^{-1}T^{-2} = [\mu]LT^{-1}L^{-1} = [\mu]T^{-1}$$

That is,

$$[\mu] = ML^{-1}T^{-1}$$

Therefore, the unit of μ is kg/(m·s). At 0℃ the coefficient of absolute viscosity of dry air is 1.716×10^{-5} kg/(m·s). The coefficient of absolute viscosity μ is also called the dynamic viscosity coefficient.

The Equation (2.2), with μ as constant, does not apply to all fluids. For a class of fluids, which includes blood, some oils, some paints and so called thixotropic fluids, μ is not constant but is a function of du/dh The derivative du/dh is a measure of the rate at which the fluid is shearing. Usually μ is expressed as (N·s)/m² or g/(cm·s). One g/(cm·s) is known as a poise.

From Newton's law of viscosity, that is, the stresses which oppose the shearing of a fluid are proportional to the rate of shear strain, the shear stress τ is given by

$$\tau = \mu \frac{\partial u}{\partial y} \qquad (2.3)$$

where μ is the coefficient of absolute viscosity and $\partial u/\partial y$ is the velocity gradient. The coefficient of viscosity μ is a property of the fluid. Fluids which obey the above law of viscosity are termed Newtonian fluids. Some fluids such as silicone oil, viscoelastic fluids, sugar syrup, tar, etc. do not obey the viscosity law given by Equation (2.3) and they are called non-Newtonian fluids.

We know that, for incompressible flows, it is possible to separate the calculation of velocity boundary layer from that of thermal boundary layer. But for compressible flows it is not possible, since the velocity and thermal boundary layers interact intimately and hence, they must be considered simultaneously. This is because, for high-speed flows (compressible flows) heating due to friction as well as temperature changes due to compressibility must be taken into account. Further, it is essential to include the effects of viscosity variation with temperature. Usually large variations of temperature are encountered in high-speed flows.

The relation $\mu(T)$ must be found by experiment. The fundamental relationship is a complex one, and that no single correlation function can be found to apply to all gases. Alternatively, the dependence of viscosity on temperature can be calculated with the aid of the method of statistical mechanics, but as of yet no completely satisfactory theory has been evolved. Also, these calculations lead to complex expressions for the function $\mu(T)$. Therefore, only semi-empirical relations appear to be the means to calculate the viscosity associated with compressible boundary layers. It is important to realize that, even though semi-empirical relations are not extremely precise, they are reasonably simple relations giving results of acceptable accuracy. For air, it is possible to use an interpolation formula based on D. M. Sutherland's theory of viscosity and express the viscosity coefficient, at temperature T, as

青等不符合式（2.3）给出的黏性定律，它们被称为非牛顿流体。

我们知道，对于不可压缩流动，可以分别计算速度边界层和热边界层。但是对于可压缩流动，则不可能。因为此时速度边界层和热边界层相互作用，密不可分，因此，必须同时加以考虑。这是因为，对于高速流动（可压缩流动），要同时考虑摩擦引起的发热和可压缩性导致的温度改变。此外，还必须考虑黏度随温度变化的影响。通常高速流动中温度变化很大。

黏度随温度变化的关系式 $\mu(T)$ 只能通过实验获得。$\mu(T)$ 的基本关系式是个复杂的式子，并且没有一个唯一的相关函数能够适用于所有气体。或者，人们可以借助统计力学方法来计算黏度随温度变化的关系，但这种方法至今还没有发展成一种完全令人满意的理论。而且，这些计算给出的是函数 $\mu(T)$ 的复杂表达式。因此，似乎只有半经验公式的方法是计算可压缩边界层黏度的方法。重要的是要认识到尽管半经验公式不是十分精确，却是能够给出足够精确结果的比较简单的关系式。对于空气这种介质，可以用一个基于萨瑟兰黏性理论的插值公式来计算黏度，即把温度 T 时的黏度表述为

$$\frac{\mu}{\mu_0} = \left(\frac{T}{T_0}\right)^{3/2} \frac{T_0 + S}{T + S}$$

where μ_0 denotes the viscosity at the reference temperature T_0, and S is a constant, which assumes the value 110K for air.

For air the Sutherland's relation can also be expressed as

式中，μ_0 是参考温度 T_0 时的黏度；S 是一个常数，对于空气介质，其取值为110K。

空气的萨瑟兰关系式也可表示为

$$\mu = 1.46 \times 10^{-6} \frac{T^{3/2}}{T+111} \quad (\text{N} \cdot \text{s}/\text{m}^2) \tag{2.4}$$

where T is in Kelvin. This equation is valid for the static pressure range of 0.01 to 100atm, which is commonly encountered in atmospheric flight. The temperature range in which this equation is valid is from 0 to 3000K. The reasons that the absolute viscosity is a function only of temperature under these conditions are that the air behaves as a perfect gas, in the sense that intermolecular forces are negligible, and that viscosity itself is a momentum transport phenomenon caused by the random molecular motion associated with thermal energy or temperature.

式中，T 以K为单位。这个关系式适用于0.01~100atm的静压变化范围，这通常是大气中飞行所遇到的压力范围。适用的温度变化范围为0~3000K。在这些条件下，因为分子间作用力可以被忽略，黏性本身只是一个与热能或温度相关的随机分子运动引起的动量交换现象，空气表现为完全气体，绝对黏度仅是温度的函数。

2. Coefficient of Kinematic Viscosity

The coefficient of kinematic viscosity is a convenient form of expressing the viscosity of a fluid. It is formed by combining the density ρ and the coefficient of absolute viscosity μ, according to the equation

2. 运动黏度

运动黏度是流体黏性的一种实用表达形式，按照下式把密度 ρ 和绝对黏度 μ 相结合来表示：

$$\nu = \frac{\mu}{\rho} \tag{2.5}$$

The coefficient of kinematic viscosity ν is expressed as m^2/s and 1cm^2/s is known as stoke.

The coefficient of kinematic viscosity is a measure of the relative magnitudes of viscosity and inertia of the fluid. Both dynamic viscosity coefficient μ and kinematic viscosity coefficient ν are functions of temperature. For liquids, μ decreases

运动黏度 ν 用单位 m^2/s 和 cm^2/s 表示。1cm^2/s 就是 1St（1斯托克斯）。

运动黏度是流体黏性和惯性相对大小的一种度量。动力黏度 μ 和运动黏度 ν 都是温度的函数。对于液体，μ 随温度的增加而减小；然而对于气体，μ 随温

with increase of temperature, whereas for gases μ increases with increase of temperature. This is one of the fundamental differences between the behavior of gases and liquids. The viscosity is practically unaffected by the pressure.

度的增加而增加。这是气体和液体表现出的一个本质不同。实际应用中黏性不受压力的影响。

2.2.5 Compressibility

2.2.5 可压缩性

The change in volume of a fluid associated with change in pressure is called compressibility. When a fluid is subjected to pressure, it gets compressed and its volume changes. The bulk modulus of elasticity is a measure of how easily the fluid may be compressed, and is defined as the ratio of pressure change to volumetric strain associated with it. The bulk modulus of elasticity, K, is given by

随压力变化流体体积发生变化的性质称为可压缩性。当流体受压力作用时，会被压缩，体积发生变化。体积弹性模量是流体被压缩容易程度的一种度量，被定义为压力变化与由其引起的体积变化率的比值。体积弹性模量 K 由下式给出：

$$K = \frac{\text{Pressure increment（压力增量）}}{\text{Volume strain（体积形变）}} = -\frac{\mathrm{d}p}{\mathrm{d}V/V}$$

This may also be expressed as

该关系式也可表示为

$$K = \lim_{\Delta v \to 0} \frac{-\Delta p}{\Delta v/v} = \frac{\mathrm{d}p}{\mathrm{d}\rho/\rho} \tag{2.6}$$

where v is specific volume. Since $\mathrm{d}\rho/\rho$ represents the relative change in density brought about by the pressure change $\mathrm{d}p$, it is apparent that the bulk modulus of elasticity is the inverse of the compressibility of the substance at a given temperature. For instance, K for water and air are approximately 2GN/m^2 and 100kN/m^2, respectively. This implies that, air is about 20000 times more compressible than water. It can be shown that, $K = a^2/\rho$, where a is the speed of sound. The compressibility plays a dominant role at high speed. Mach number Ma (defined as the ratio of local flow velocity to local speed of sound) is a convenient non-dimensional parameter used in the study of compressible flows. Based on Ma the flow is divided into the following regimes. When $Ma < 1$ the flow is called subsonic, when $Ma \approx 1$ the flow is termed transonic flow, Ma from 1 to 5 is called

式中，v 是比容，$v = 1/\rho$。由于 $\mathrm{d}\rho/\rho$ 表示由压力变化 $\mathrm{d}p$ 引起的密度相对变化，显然体积弹性模量是在给定温度下物质可压缩性的倒数。例如，对于水和空气，K 大约分别是 2GN/m^2 和 100kN/m^2，这表示空气比水的可压缩性约大 20000 倍。推导可得，$K = a^2/\rho$，其中 a 是声速。流体在高速流动时可压缩性起主导作用。在可压缩流动的研究中马赫数 Ma（定义为当地流速和当地声速之比）是一个实用的无量纲参数。基于马赫数 Ma，流动被分为如下范畴：当 $Ma < 1$ 时，流动称为亚声速；当 $Ma \approx 1$ 时，流动称为跨声速；当 Ma 在 1 到 5

supersonic regime, and $Ma > 5$ is referred to as hypersonic regime. When flow Mach number is less than 0.3, the compressibility effects are negligibly small, and hence the flow is called incompressible. For incompressible flows, density change associated with velocity is neglected and the density is treated as invariant.

2.3 Thermodynamic Properties of Fluids

We know from thermodynamics that, heat is energy in transition. Therefore, heat has the same dimensions as energy, and is measured in units of joule (J).

2.3.1 Specific Heat

The inherent thermal properties of a flowing gas become important when the Mach number is greater than 0.5, that is, when the energy equation needs to be considered. The specific heat is one such quantity. The specific heat is defined as the amount of heat required to raise the temperature of a unit mass of a medium by 1K. The value of the specific heat depends on the type of process involved in raising the temperature of the unit mass. Usually constant volume process and constant pressure process are used for evaluating specific heat. The specific heats at constant volume and constant pressure processes, respectively, are designated by c_v and c_p. The definitions of these quantities are the following.

$$c_v \equiv \left(\frac{\partial u}{\partial T}\right)_v \tag{2.7}$$

where u is internal energy per unit mass of the fluid, which is a measure of the potential and more particularly the kinetic energy of the molecules comprising the gas. The specific heat at constant volume c_v is a measure of

2.3 流体的热力学性质

由热力学可知，热是传递的能量。因此，热量和能量具有相同的量纲，用单位焦耳（J）来度量。

2.3.1 比热

当马赫数大于 0.5，即需要考虑能量方程时，流动气体的固有热力学性质就变得很重要，比热就是这样一个重要的量，它定义为使单位质量介质温度升高 1K 所需的热量。比热值依赖于单位质量物质温度升高过程的类型，通常用定容过程和定压过程来求比热值。定容和定压过程的比热分别用 c_v 和 c_p 表示，这些量的定义如下。

式中，u 是单位质量流体的内能，它是组成气体的分子势能以及动能的一种度量。定容比热 c_v 是气体分子携带能量能力的一种度

the energy-carrying capacity of the gas molecules. For dry air at normal temperature, $c_v = 717.5 \text{J}/(\text{kg}\cdot\text{K})$.

The specific heat at constant pressure is defined as

$$c_p \equiv \left(\frac{\partial h}{\partial T}\right)_p \tag{2.8}$$

where $h = u + pv$, the sum of internal energy and flow energy is known as the enthalpy or total heat constant per unit mass of fluid. The specific heat at constant pressure c_p is a measure of the ability of the gas to do external work in addition to possessing internal energy. Therefore, c_p is always greater than c_v. For dry air at normal temperature, $c_p = 1004.5 \text{J}/(\text{kg}\cdot\text{K})$.

Note: It is essential to understand what is meant by normal temperature. For gases, up to certain temperature, the specific heats will be constant and independent of temperature. Up to this temperature the gas is termed perfect, implying that c_p, c_v and their ration γ are constants, and independent of temperature. But for temperatures above this limiting value, c_p, c_v will become functions of T, and the gas will cease to be perfect. For instance, air will behave as perfect gas up to 500K. The temperature below this limiting level is referred to as normal temperature.

2.3.2 The Ratio of Specific Heats

The ratio of specific heats,

$$\gamma = \frac{c_p}{c_v} \tag{2.9}$$

is an important parameter in the study of high-speed flows. This is a measure of the relative internal complexity of the molecules of the gas. It has been determined from kinetic theory of gases that, the ratio of specific heats can be related to the number of degrees of freedom, n, of the gas molecules by the relation:

$$\gamma = \frac{n+2}{n} \tag{2.10}$$

At normal temperatures, there are six degrees of freedom, namely three translational and three rotational degrees of freedom, for diatomic gas molecules. For nitrogen, which is a diatomic gas, $n=5$ since one of the rotational degrees of freedom is negligibly small in comparison with the other two. Therefore,	在常温下，对于双原子气体分子，存在6个自由度，即3个平动和3个转动。对于氮气——一种双原子气体，因为一个转动自由度相对于另外两个转动自由度小到可以忽略，$n=5$。因此，

$$\gamma = 7/5 = 1.4$$

Monatomic gases, such as helium, have 3 translational degrees of freedom only, and therefore,	单原子气体，例如氦气，仅有3个平动自由度，因此

$$\gamma = 5/3 = 1.67$$

This value of 1.67 is the upper limit of the values which the ratio of specific heats γ can take. In general γ varies from 1 to 1.67, that is,	这个1.67的数值是所有比热比γ可取值的上限。通常γ在1到1.67之间变化，即

$$1 \leqslant \gamma \leqslant 1.67$$

The specific heats of a gas are related to the gas constant R. For a perfect gas this relation is	气体的比热和气体常数R有关。对于完全气体，此关系式为

$$R = c_p - c_v$$

2.3.3 Thermal Conductivity of Air / 2.3.3 空气的导热性

At high speed, heat transfer from vehicles becomes significant. For example, re-entry vehicles encounter an extreme situation where ablative shields are necessary to ensure protection of the vehicle during its passage through the atmosphere. The heat transfer from a vehicle depends on the thermal conductivity k of air. Therefore, a method to evaluate k is also essential. For this case, a relation similar to Sutherland's law for viscosity is found to be useful, and it is	在高速飞行时，飞行器的热交换非常重要。例如，再入飞行器会遇到一种极端的情况，飞行器穿越大气时，必须采用可烧蚀的防护罩来保护飞行器。飞行器换热与空气的热导率k有关。因此，需要一种用于估算k的方法。对于上述例子，可用一个类似于萨瑟兰黏性定律的关系，即

$$k = 1.99 \times 10^{-3} \frac{T^{3/2}}{T+112} \quad [J/(s \cdot m \cdot K)]$$

where T is temperature in Kelvin. The pressure and temperature ranges in which this equation is applicable are 0.01 to 100atm and 0 to 2000K,	式中，T是以K为单位的温度。方程适用的压力和温度范围分别是0.01～100atm和0～2000K。和

respectively. For the same reason given for viscosity relation, the thermal conductivity also depends only on temperature.

2.4 Surface Tension

Liquids behave as if their free surfaces were perfectly flexible membranes having a constant tension σ per unit width. This tension is called the surface tension. It is important to note that, this is neither a force nor a stress but a force per unit length. The value of surface tension depends on the nature of the fluid, the nature of the surface of the substance with which it is in contact, the temperature and pressure.

Consider a plane material membrane, possessing the property of constant tension σ per unit length. Let the membrane has a straight edge of length l. The force required to hold the edge stationary is

$$p = \sigma l \tag{2.11}$$

Now, suppose that the edge is pulled so that it is displaced normal to itself by a distance x in the plane of the membrane. The work done, F, in stretching the membrane is given by

$$F = \sigma l x = \sigma A \tag{2.12}$$

where A is the increase in the area of the membrane. It is seen that, σ is the free energy of the membrane per unit area. The important point to be noted here is that, if the energy of a surface is proportional to its area, then it will behave exactly as if it was a membrane with a constant tension per unit length, and this is totally independent of the mechanism by which the energy is stored. Thus, the existence of surface tension, at the boundary between two substances, is a manifestation of the fact that the stored energy contains a term proportional to the area of the surface. This energy is attributable to molecular attractions.

An associated effect of surface tension is the capillary deflection of liquids in small tubes. Examine the level of water and mercury in capillaries, shown in Figure 2.3.

与表面张力相关的一个作用是细小管中液体的毛细偏移，如图 2.3 所示，观察毛细管中水和汞的高度。

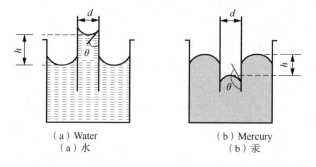

(a) Water　　　　　　(b) Mercury
(a) 水　　　　　　　　(b) 汞

Figure 2.3　Capillary effect of water and mercury

图 2.3　水和汞的毛细效应

When a glass tube is inserted into a beaker of water, the water will rise in the tube and display a concave meniscus, as shown in Figure 2.3(a). The deviation of water level h in the tube from that in the beaker can be shown to be

当一烧杯水中插入一个玻璃管时，水会在管内上升并显示为一个凹的半月状，如图 2.3（a）所示。管内水面偏离烧杯中液面的高度 h 可表示为

$$h \propto \frac{\sigma}{d}\cos\theta \tag{2.13}$$

where θ is the angle between the tangent to the water surface and the glass surface. In other words, a liquid such as water or alcohol, which wets the glass surface, makes an acute angle with the solid, and the level of free surface inside the tube will be higher than that outside. This is termed capillary action. However, when wetting does not occur, as in the case of mercury in glass, the angle of contact is obtuse, and the level of free surface inside the tube is depressed, as shown in Figure 2.3(b).

式中，θ 是水面的切线和玻璃表面之间的夹角。换句话说，像水或者酒精之类能够使玻璃表面浸润的液体，会与固体之间形成一个锐角，于是管内自由表面的高度将比外面的高一些，这称为毛细作用。然而，当不发生浸润时，如汞在玻璃管中的情况，接触角是钝角，则管内自由表面的高度降低，如图 2.3（b）所示。

Another important effect of surface tension is that, a long cylinder of liquid, at rest or in motion, with a free surface is unstable and breaks up into parts, which then assumes an approximately spherical shape. This is the mechanism of the breakup of liquid jets into droplets.

表面张力的另一个重要作用是，一个处于静止或运动状态且有着自由表面的长液柱是不稳定的，会破碎成近似球形的碎片，这就是液体射流破碎成液滴的机理。

2.5 Summary

This chapter mainly denotes the basic properties of fluids, including pressure, temperature, density, viscosity and compressibility. Both the properties of gases and liquids are introduced. In addition, the thermodynamic properties of fluids, including specific heat and thermo conductivity are given. The surface tension of fluids are also briefly illustrated.

2.6 Exercises

Problem 2.1 Compare the column heights of water and mercury corresponding to a pressure of 100kPa. Express your answer in meters.

[Ans: 10.194m of water, 0.750m of mercury]

Problem 2.2 A large airship of volume 90000m^3 contains helium under standard atmospheric conditions. Determine the density and total weight of the helium.

[Ans: $0.1686\text{kg/m}^3, 1.4886 \times 10^5 \text{N}$]

Problem 2.3 The lower-half of a 5m high circular cylinder container is filled with water and the upper-half with an oil of specific gravity 0.8. Determine the pressure difference between the top and bottom of the cylinder.

[Ans: 44.145kPa]

Problem 2.4 A gas is contained in a vertical, frictionless piston-cylinder device. The piston has a cross-sectional area of 30cm^2 and a mass of 3kg. A compressed spring above the piston exerts a force of 50N on the piston. Calculate the pressure inside the cylinder. Take atmospheric pressure to be 100kPa.

[Ans: 126.476kPa]

Problem 2.5 A right circular conical tank of height

2.5 小结

本章主要论述了流体的基本特性，包括压力、温度、密度、黏性和可压缩性，并对气体和液体的特性均进行了介绍。此外，本章还给出了流体的热力学特性，包括比热和导热性，并简单介绍了流体的表面张力。

2.6 习题

题 2.1 比较 100kPa 压力对应的水柱和汞柱高度，以米为单位给出结果。

【答：水柱高 10.194m，汞柱高 0.750m】

题 2.2 一个体积为 90000m^3 的飞艇，充满标准大气压下的氦气。求该氦气的密度和总重量。

【答：$0.1686\text{kg/m}^3, 1.4886 \times 10^5 \text{N}$】

题 2.3 一个 5m 高圆柱形容器的下半部充满水而上半部是比重为 0.8 的油，求容器顶部和底部之间的压力差。

【答：44.145kPa】

题 2.4 一个无摩擦的竖直活塞缸充满气体，活塞截面为 30cm^2，质量为 3kg，活塞上的压缩弹簧施加 50N 的力给活塞，计算缸中压力，假设大气压为 100kPa。

【答：126.476kPa】

题 2.5 一个高 2m、基底直径 1m

2m and base diameter 1m is filled with 0.5m³ of water. Determine the height of the free surface of water from the tank base.

[Ans: 1.291m]

的正圆锥容器装有 0.5m³ 的水，求容器底到水面的高度。

【答：1.291m】

Chapter 3 Fluid Statics
第 3 章 流体静力学

3.1 Introduction

In Chapter 1, we saw that the study of fluid behavior can be divided into statics, kinematics and dynamics. If all elements (an element is an infinitely large number of molecules) are at rest with respect to a reference coordinate system, the fluid is considered to be static. In some cases, the fluid elements may be at rest with respect to each other or with respect to the boundaries enclosing them, but not with respect to a reference coordinate system. In such cases the fluid elements and the container enclosing them may be moving with a uniformly accelerated or decelerated motion. For these cases also, the law of hydrostatics, taking into account the effects of uniform acceleration, can still be applied. Thus, when a fluid is either at rest or in relative equilibrium, the fluid elements are not subjected to any shearing stress. This chapter mainly introduces the properties and mechanism of fluids at rest or in relative equilibrium.

3.2 Scalar, Vector and Tensor Quantities

Before we enter the field of fluid mechanics, it will be useful to classify certain types of quantities which are essential for the study of the subject. Some of the very useful quantities necessary for fluid mechanics are scalar, vector, and tensor quantities.

Scalar quantities require only a magnitude to be

3.1 引言

第 1 章中我们看到对流体行为的研究可以分为流体静力学、流体运动学和流体动力学三种。如果流体中所有微元（一个微元是无限多数量的分子）相对于一个参考坐标系是静止的，这个流体就被认为是静止的。在某些情况下，流体微元可能相互之间是静止的或相对于包围液体的边界是静止的，但相对于参考坐标系不是静止的。在这样的情况下，流体微元和盛装流体的容器可能一起做等加速或等减速运动。对于这类情况，如果考虑等加速作用，流体静力学规律仍然适用。因此，当流体是静止的或处于相对平衡状态时，流体微元不受任何剪切力作用。本章主要讨论流体处于静止或平衡状态的特性及平衡规律。

3.2 标量、矢量和张量

在进入流体力学领域学习之前，对这门学科学习中必须要用到的量进行一定的分类是很有帮助的。一些非常有用的流体力学基本量是标量、矢量和张量。

标量只需指定一个数值大小

specified for a complete description. For example, temperature is a scalar quantity.

Vector quantities require, in addition to magnitude, a complete directional specification for their description. Usually three values associated with orthogonal directions are used to specify a vector. These quantities are called scalar components of a vector. For example, velocity is a vector.

Tensor quantities require the specification of nine or more scalar components for a complete description. For example, stress, strain, and mass moment of inertia are tensor quantities.

A field is a continuous distribution of a scalar, vector, or a tensor quantity described by continuous functions of space coordinates and time. For example, the scalar quantity temperature at all points in a field may be expressed mathematically as $T(x, y, z, t)$. In a similar fashion, a vector field like velocity may also be described mathematically as $V(x, y, z, t)$. However, normally three scalar fields are employed to designate a vector field. Thus, the velocity field may be expressed as

就能完全被描述。例如，温度是一个标量。

矢量，除了数值大小，还需要给定一个完整的方向定义才能完全被描述。通常用三个数值结合三个正交方向来确定一个矢量。这三个数值被称为矢量的标量分量。例如，速度是一个矢量。

张量的完整描述需要 9 个或更多个标量分量。例如，应力、应变和质量惯性矩是张量。

场是由空间坐标和时间的连续函数表示的标量、矢量或张量的连续分布。例如，场中所有点的温度标量的数学描述为 $T(x, y, z, t)$。同样，一个矢量场例如速度场的数学表达形式为 $V(x, y, z, t)$。通常一个矢量场用三个标量场表示，因此，速度场可以表示为

$$V_x = f(x, y, z, t) \quad (3.1a)$$
$$V_y = g(x, y, z, t) \quad (3.1b)$$
$$V_z = h(x, y, z, t) \quad (3.1c)$$

where V_x, V_y, V_z are the velocity components along x, y, and z directions, respectively. Likewise, the tensor fields may be designated mathematically by employing nine or more scalar fields.

式中，V_x、V_y、V_z 分别是沿 x 方向、y 方向、z 方向的速度分量。同样，张量场也可以用 9 个或更多个标量场以数学方式表示出来。

3.3 Body and Surface Forces

The forces coming across in continuum fluid mechanics may broadly be divided into body forces and surface forces. All external forces acting on any material, which are developed without physical contact, are called body forces. Gravitational force, the effect of the earth on a mass manifesting itself as

3.3 体积力和表面力

连续介质流体力学中涉及的力可以大体上分为体积力和表面力两种。所有施加在任意质点上，不由物理接触产生的外部作用力称为体积力。重力由地球作用在物体上，分布到各质点，方向指向地

a force distribution throughout the material, directed towards the earth's centre, is a body force. Body forces are usually expressed per unit mass of the material acted on. All forces exerted on a boundary by its surroundings through direct contact are termed as surface forces, for example, pressure.

3.4 Forces in Stationary Fluids

Consider an infinitesimal prismatic element in a stationary fluid, as shown in Figure 3.1. The triangular prismatic element considered has dimensions δx, δy and δz along x, y, and z directions, respectively. Assume that the dimension δy along y direction is one unit of length. That is $\delta y = 1$. The gravity is the only force acting on the fluid element. There are no shear forces acting on the element, since a fluid element cannot withstand a shear stress, and a stationary fluid should necessarily be completely free of shear stress.

心，便是一种体积力。体积力通常表示为单位质量上的作用力。所有由周围介质通过直接接触作用在物体边界上的力都称为表面力，例如压力。

3.4 静止流体中的力

在静止流体中取一个无限小的楔形微元，如图 3.1 所示。被选取的三角楔形微元沿 x、y、z 方向的边长分别为 δx、δy、δz。假设 y 方向边长 δy 为单位长度，即 $\delta y = 1$。重力是流体微元上唯一的作用力，由于流体微元不能承受剪切应力，因此流体微元上不受剪切力作用，所以静止流体上应完全不受剪切应力作用。

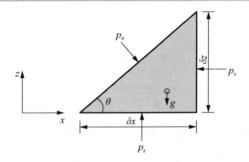

Figure 3.1 A stationary fluid element

图 3.1 静止流体微元

The pressures acting on the faces are shown as p_x, p_z and p_n. The pressure, p, may be defined as "the force per unit area which acts normal to the surface of any object which is immersed in a fluid". For equilibrium, the net force acting on the fluid element along x and z directions must be zero. Therefore,

作用在各表面上的压力为 p_x、p_z 和 p_n。压力 p 可定义为垂直作用在浸没于流体中的任何物体表面单位面积上的作用力。为保持平衡，沿 x 和 z 方向作用于流体微元的合力一定为零。因此，

$$p_x(\delta z \cdot 1) - p_n \frac{\delta z}{\sin \theta} \sin \theta = 0 \tag{3.2}$$

$$p_z(\delta x \cdot 1) - p_n \frac{\delta z}{\cos\theta}\cos\theta - \frac{1}{2}(\delta x \cdot \delta z \cdot 1)g = 0 \tag{3.3}$$

Now, letting the size of the element shrink to zero, we see from Equations (3.2) and (3.3) that	现在，让微元尺寸趋近于零，由式（3.2）和式（3.3）可见，

$$p_x = p_y = p_z = p \tag{3.4}$$

That is, the pressure in a stationary fluid is equal in all directions.	即静止流体中压力在各个方向上是相等的。

3.5 Pressure Force on a Fluid Element / 3.5 流体微元上的压力合力

Consider a small fluid element of length δx, δy and δz along x, y, and z directions, respectively, in a stationary fluid, as shown in Figure 3.2. The corner of fluid element close to the origin is taken as the position (x, y, z).	在静止流体中取一个沿 x、y、z 方向边长分别为 δx、δy 和 δz 的流体微元，如图 3.2 所示。设流体微元接近原点的一角为位置 (x,y,z)。

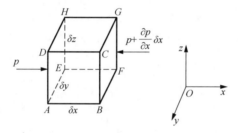

Figure 3.2　Pressure force on a fluid element

图 3.2　流体微元上的压力

The pressure force acting on the fluid element along the x-direction is given by	沿 x 方向作用在流体微元上的压力可写为

$$dF_{p,x} = p(\delta y \cdot \delta z) - \left(p + \frac{\partial p}{\partial x}\delta x\right)(\delta y \cdot \delta z) = -\frac{\partial p}{\partial x}(\delta x \cdot \delta y \cdot \delta z) \tag{3.5}$$

Similarly, the pressure force acting on the fluid element along the y and z directions can be expressed respectively as	同样，沿 y 和 z 方向作用在流体微元上的压力可分别表示为

$$dF_{p,y} = -\frac{\partial p}{\partial y}(\delta x \cdot \delta y \cdot \delta z) \tag{3.6}$$

$$dF_{p,z} = -\frac{\partial p}{\partial z}(\delta x \cdot \delta y \cdot \delta z) \tag{3.7}$$

By combining the pressure force components given by Equations (3.5) to (3.7), the net force acting	合并式（3.5）～式（3.7）中压力项，作用在流体微元上的合力

on the fluid element can be written as | 可写为

$$dF_p = -\left(\frac{\partial p}{\partial x}i + \frac{\partial p}{\partial y}j + \frac{\partial p}{\partial z}k\right)(\delta x \cdot \delta y \cdot \delta z) \quad (3.8)$$

where i, j and k are the unit vectors along x, y, and z directions, respectively. Then the net force per unit volume is | 式中，i、j 和 k 分别是 x、y 和 z 方向上的单位矢量。那么单位体积上的合力为

$$\frac{dF}{dxdydz} = f = -\left(\frac{\partial p}{\partial x}i + \frac{\partial p}{\partial y}j + \frac{\partial p}{\partial z}k\right) \quad (3.9)$$

If cylindrical coordinates rather than Cartesian coordinates were used, f in Equation (3.9) would have taken a form different from the one given above. However, all such formulations have identically the same physical meaning which is independent of the coordinate system used for evaluation purpose. Hence, Equation (3.9) can also be written as | 如果使用柱坐标而不是笛卡儿坐标，式（3.9）中 f 将具有和上述给定形式不同的形式。然而，所有这些形式都有着同样的不依赖于所采用坐标系的物理意义。因此式（3.9）也可以写为

$$f = -\text{grad} p \quad \text{or（或）} \quad f = -\nabla p \quad (3.10)$$

where the operator ∇ is called the gradient operator and has a form dependent on the coordinate system used. In Cartesian coordinates, | 式中，∇ 为梯度算子，其形式依赖于所使用的坐标系。在笛卡儿坐标系中，

$$\text{grad} \equiv \frac{\partial}{\partial x}i + \frac{\partial}{\partial y}j + \frac{\partial}{\partial z}k \quad (3.11)$$

3.6 Basic Equations of Fluid Statics

3.6 流体静力学基本方程

Consider the element of sides δx, δy and δz of a stationary fluid, as shown in Figure 3.3. It is at rest under the action of pressure forces and gravity force. | 取边长为 δx、δy 和 δz 的静止流体微元，如图 3.3 所示，该微元在压力和重力作用下处于静止状态。

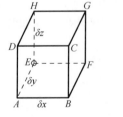

Figure 3.3 A fluid element in equilibrium

图 3.3 平衡状态下的流体微元

The pressure force acting on it is obtained from Equation (3.8), as

作用在微元上的压力可由式（3.8）求得。

$$dF_p = -\left(\frac{\partial p}{\partial x}i + \frac{\partial p}{\partial y}j + \frac{\partial p}{\partial z}k\right)(\delta x \cdot \delta y \cdot \delta z) = -(\nabla p)\delta V \tag{3.12}$$

where $\delta V = \delta x \cdot \delta y \cdot \delta z$ is the volume of the fluid element. The gravity force acting on the element is

式中，$\delta V = \delta x \cdot \delta y \cdot \delta z$ 是流体微元的体积。作用在微元上的重力为

$$dF_g = \rho g \delta V \tag{3.13}$$

For equilibrium, from Equations (3.12) and (3.13), we have

对于平衡状态，由式（3.12）和式（3.13），有

$$-\nabla p + \rho g = 0$$

or

或

$$\nabla p = \rho g \tag{3.14}$$

Equation (3.14) is the basic equation of fluid statics.

式（3.14）是流体静力学基本方程。

If g is taken as acting in the negative z-direction, that is, $g = -gk$, then the three components of Equation (3.14) are

如果 g 被认为是作用在 z 的负方向，即 $g = -gk$，那么式（3.14）三个方向的分量为

$$\frac{\partial p}{\partial x} = 0 \tag{3.15a}$$

$$\frac{\partial p}{\partial y} = 0 \tag{3.15b}$$

$$\frac{\partial p}{\partial z} = -\rho g \tag{3.15c}$$

From Equations (3.15a) - (3.15c), it is seen that pressure in a stationary fluid can vary only in the z-direction, which has been selected as the direction of gravity. In other words, the pressure in a stationary fluid varies only in the vertical direction, and is constant in any horizontal plane. At this stage, it is important to note that in the above formulations it is assumed that the free surface of a liquid at rest (or the interface between a liquid and a gas or between two immiscible liquids) is at right angles to the direction of gravity.

由式（3.15a）～式（3.15c）可见，静止流体中压力只沿被设定为重力方向的 z 方向变化。换句话说，静止流体中压力只沿垂直方向变化，而在任何水平方向上都是恒定的。这里尤其要指出的是，上述公式中假定的是静止液体的自由表面（或液体和气体之间交界面或两种不相混合的液体之间交界面）和重力方向垂直。

3.6.1 Hydrostatic Pressure Distribution

3.6.1 流体静压分布

By using ordinary derivative and realizing that the

利用常微分和压力只沿 z 方

pressure varies only in the z-direction and is not a function of x and y, Equation (3.15) may be expressed as	向变化，而不是 x 和 y 的函数这一事实，式（3.15）可表示为

$$\frac{dp}{dz} = -\rho g \qquad (3.16)$$

For incompressible fluids, that is, for fluids with constant density (ρ), Equation (3.16) can be integrated between any desired limits to evaluate the pressure distribution in the static fluid under consideration. Choosing the subscript "0" to represent conditions at the free surface, integrate Equation (3.16) from any position z, where the pressure is p, to position z_0, where the pressure is atmospheric and denoted as p_{atm}. Thus, we have	对于不可压缩流体，即密度（ρ）恒定的流体，可在任何期望的位置对式（3.16）积分来计算所选取静止流体中的压力分布。用下标"0"来代表自由表面的条件，从压力为 p 的任一位置 z 开始到压力为大气压 p_{atm} 的位置 z_0 对式（3.16）积分，于是，有

$$\int_z^{z_0} -\rho g dz = \int_p^{p_0} dp$$

Since ρ and g are constant, the above equation readily gets integrated to yield	由于 ρ 和 g 是常数，上述方程可直接积分得到

$$p_{atm} - p = -\rho g(z_0 - z)$$
$$p = p_{atm} + \rho g(z_0 - z) \qquad (3.17)$$

From Equation (3.17) it is seen that, in stationary fluids, the pressure increases linearly with depth (negative z). This linear pressure distribution is called *hydrostatic pressure distribution*. Usually, the term $(p - p_{atm})$, that is, the pressure above atmospheric pressure, is known as the gauge pressure, and is denoted by p_g. So,	式（3.17）表明，静止流体中，压力随深度（负 z 方向）线性增加，这个线性的压力分布称为**流体静压分布**。 通常，$(p - p_{atm})$ 项，即大气压以上的压力，称为表压力，用 p_g 表示。因此，

$$p_g = \rho g(z_0 - z)$$

(z_0–z) in the above equation is the depth "h" below the free surface. Therefore,	上式中 ($z_0 - z$) 是自由表面下的深度"h"，因此

$$p_g = \rho g h \qquad (3.18)$$

In all engineering flow problems, the p to be measured by pressure gauges is above or below that of atmosphere. Therefore, in engineering work the gauge pressure p_g can be negative, with a maximum possible negative value equal to $-p_{atm}$. The hydrostatic pressure distribution, see Equation (3.18), also holds for moving fluids,	在所有工程流动问题中，p 是由压力表测量的高于或低于大气压的压力。因此，在工程实际中表压力 p_g 可以是负的，可能的最低负压力值等于 $-p_{atm}$。 对于流动流体，如果没有垂直于流动方向的加速度，式（3.18）

provided there is no acceleration in the direction normal to the flow. This finds very good application in manometry.

3.6.2 Measurement of Pressures

The hydrostatic pressure distribution given by Equation (3.16) is directly useful for measurement of pressure in both compressible and incompressible flows. Let us see the solution of this equation for both the flows.

1. Constant Density Solution (Incompressible Fluid)

Equation (3.16) may be rearranged as

$$dp = \rho g dh$$

To compare the pressures p_1 and p_2 at two points in a fluid at heights h_1 and h_2, respectively, this equation may be integrated between condition 1 and condition 2 as

$$\int_1^2 dp = \int_1^2 \rho g dh$$

In the case where ρ and g are constants this may be integrated as

$$p_2 - p_1 = \rho g(h_2 - h_1)$$

If $(h_2 - h_1)$ is positive (point 2 above point 1), $(p_1 - p_2)$ is also positive $(p_2 < p_1)$. Thus, pressure decreases with height. Further, $\rho g(h_2 - h_1)$ is the weight of a column of fluid of unit area and height $(h_2 - h_1)$. It may thus be deduced that "the pressure at any point in a fluid at rest is equal to the weight of all fluid contained in a column of unit area above that point". It is important to note that most liquids have a density which may be regarded as constant, but this is not true for gases, except over very small height differences.

From the above discussion it is evident that, a pressure difference used to support a column of liquid of known density, can be measured by the height of that column. The mercury barometer is a well-known example of this. Standard atmospheric pressure of 101325Pa is approximately equal to 760mm of mercury column height [$\rho g h$ =101325Pa, h=101325/(13.6×10³×9.81)=0.7595m]. A practical application of this principle is the U-tube, which forms the basis of several pressure-measuring instruments. Any pressure measuring instrument is called a manometer. The simplest form of manometer is a tube bent into the form of a "U" and partly filled with liquid, as shown in Figure 3.4.

The two limbs are connected to the two points between which the pressure difference is to be measured. Equation (3.18) gives

测量支撑流体柱所需的压力差。汞柱气压计就是众所周知的一个例子。标准大气压力约为760mmHg [$\rho g h$ = 101325Pa, h = 101325/(13.6×10³×9.81) = 0.7595 m]。几种压力测量装置的基本组成部分——U 形管，就是这个原理的一个实际应用。所有压力测量装置都称为压力计，最简单的压力计是被弯成 U 形填充一部分液体的管，如图 3.4 所示。

两个支管分别连接被测压力差的两点，由式（3.18）有

$$p_1 - p_2 = -\rho g(h_2 - h_1)$$

or 或

$$\Delta p = \rho g \Delta h$$

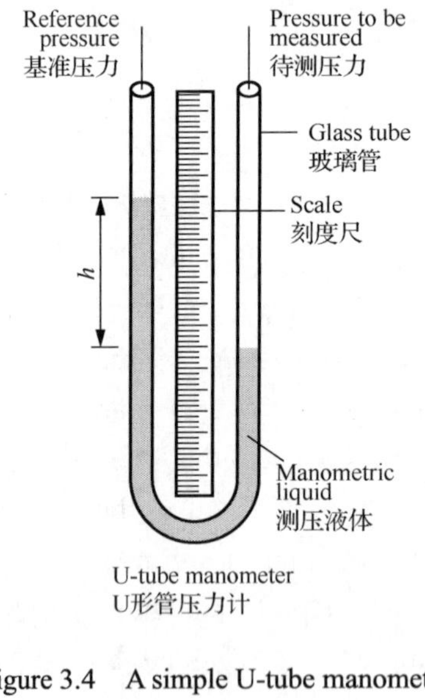

Figure 3.4　A simple U-tube manometer

图 3.4　一个简单的 U 形管压力计

Thus, by measuring Δh and using the known value of ρ it is possible to calculate the pressure difference Δp, taking care to use consistent units.

因此，如果使用统一单位，通过测量 Δh 并利用已知的 ρ 值，就能够计算出压力差 Δp。

For situations where it is necessary to measure pressure differences corresponding to less than 1mm of water or less, the simple U-tube is not sufficiently sensitive and various other forms of manometer have been developed to measure these small differences. Most of these are based on the U-tube and differ only in the method used to give the increased sensitivity. Two such popular methods are:

(1) Making the fluid displacement to be measured greater than the displacement in a simple U-tube. A common example to this is the inclined tube manometer. The fluid displacement is increased by the factor cos θ, θ being the angle of slope to the horizontal.

(2) Using some optical arrangement to magnify the actual measurement, for example Betz manometer.

The details of construction and operating principle can be found in the book *Instrumentation Measurement and Experiments in Fluid* by E. Rathakrishnan[1]. Many instruments use alcohol or similar fluids with densities less than that of water to obtain larger displacement.

对于需要测量小于 1mm H_2O（1mm H_2O = 9.80665Pa）或更小压力差的情况，简单的 U 形管是不够灵敏的。为测量很小的压力差，人们开发出了各种其他形式的压力计。这些压力计大多数还是以 U 形管为基础，只是所使用的提高精度的方法有所不同。两个比较常用的方法是：

（1）使被测流体液面差比简单 U 形管中液面差更大，常见的例子是倾斜管压力计，液面差通过 $\cos\theta$ 因子被增大，θ 是相对于水平面的倾角。

（2）用一些光学方法把实际测量值放大，例如贝茨压力计。

贝茨压力计的结构和工作原理可详见 E. Rathakrishnan 的 *Instrumentation, Measurements and Experiments in Fluids* 一书[1]。此外，许多压力计还用酒精或类似密度小于水的流体来得到更大的液面差。

2. Variable Density Solution (Compressible Fluids)

When the density is variable and not constant, the functional dependent of other flow parameters should be known for solving Equation (3.16). For perfect gases we can make use of the equation of state for this. For perfect gases, the equation of state relating the density to pressure and temperature is

2. 变密度解（可压缩流体）

当密度是变量而不是常数时，求解式（3.16）需要知道密度与其他流动参数的函数关系。对于完全气体，可利用状态方程。联系密度和压力以及温度之间关系的完全气体状态方程为

$$\rho = \frac{p}{RT}$$

With the expression for ρ, Equation (3.16) can be expressed as

根据 ρ 的上述表达式，式（3.16）可表示为

$$\frac{\mathrm{d}p}{p} = -\frac{g}{RT}\mathrm{d}z \qquad (3.19)$$

The pressure variation in a static compressible fluid can be determined from Equation (3.19) if the temperature as a function of z is known. Equation (3.19), which may be called the "aerostatic equation", has an important application in the determination of the properties of the atmosphere.

3.6.3 Units and Scales of Pressure Measurement

The fundamental problem of fluid statics is the determination of the static pressure in a fluid. Based on Equation (3.19) various devices for measuring static pressure have been developed. These devices are generally known as manometers. Pressure may be expressed with reference to any arbitrary datum. The usual references are *absolute zero* (complete vacuum) and *local atmospheric pressure*. When a pressure is expressed as a difference between its value and a complete vacuum, it is called an *absolute pressure*. When it is expressed as a difference between its value and the local atmospheric pressure, it is termed as *gauge pressure*. These two scales for pressure measurement can be expressed as follows:

$$p_{abs} = p - 0 = p \qquad (3.20a)$$
$$p_{gauge} = p - p_{atm} \qquad (3.20b)$$

If the pressure p is less than the atmospheric pressure, the gauge pressure has a negative value. This is called a vacuum pressure.

3.7 The Atmosphere

The atmosphere may be regarded as an expanse of fluid (air) substantially at rest. Hydrostatic theory may be used to calculate its macroscopic properties. The atmosphere is a mixture of gases of which nitrogen and oxygen are the main constituents. It also contains small amounts of other gases, including hydrogen and helium and the rare inert gases argon, krypton, neon, etc. Over the range of altitudes involved in conventional aerodynamics, the properties of the constituents vary little, and the atmosphere may be regarded as a homogeneous gas of uniform composition.

It is well established that the atmosphere may conveniently be divided into two distinct continuous regions. The lower of these regions is called the *troposphere*, and it is found that the temperature within the troposphere decreases approximately linearly with height. The upper region is the *stratosphere* which in the temperature remains almost constant with height. The supposed boundary between the two regions is termed the *tropopause*. The sharp distinction between the two, implied above, does not exist in reality, but one merges gradually into the other. Nevertheless, this distinction represents a useful convention for the purpose of calculation.

3.7.1 The International Standard Atmosphere

The performance of a flying vehicle is dependent on the physical properties, like density and temperature, of the air in which it flies. Therefore, it is desirable that comparison between flying vehicles should be based on similar atmospheric conditions. To assist aircraft designs and operators in this, agreement has been reached as an International Standard Atmosphere (ISA) which intends to approximate to the atmospheric conditions prevailing for most of the year in temperate latitudes, for example, Europe and North America. The ISA is defined by the pressure and temperature at mean sea level, and the variation of temperature with altitude.

3.7.2 Calculations on the Stratosphere

It is well-known that, in the stratosphere the

temperature does not vary with height. Therefore, we need to obtain expression only for the variation of pressure and density with altitude. For this purpose, we can make use of hydrostatic pressure distribution given by Equation (3.16) and the perfect gas equation of state given by Equation (1.2).

Thus, for our calculations on the stratosphere, we have the equations

$$\frac{\mathrm{d}p}{\mathrm{d}h} = -\rho g$$

$$p = \rho RT$$

$$T = T_s$$

where T_s is the constant temperature in the stratosphere. The equation of state can be expressed as

$$\rho g = \frac{gp}{RT_s}$$

Substituting this into hydrostatic equation, we obtain

$$\frac{\mathrm{d}p}{\mathrm{d}h} = -\frac{gp}{RT} = -\frac{gp}{RT_s}$$

or

$$\frac{\mathrm{d}p}{p} = -\frac{g}{RT_s}\mathrm{d}h$$

Integrating this between heights h_1 and h_2 ($h_2 > h_1$, as h here is the altitude) and the corresponding pressures p_1 and p_2, we get

$$[\ln p]_{p_1}^{p_2} = -\frac{g}{RT_s}[h]_{h_1}^{h_2}$$

or

$$\ln\left(\frac{p_2}{p_1}\right) = -\frac{g}{RT_s}(h_2 - h_1)$$

That is

$$\frac{p_2}{p_1} = \exp\left[\frac{g(h_1 - h_2)}{RT_s}\right] \quad (3.21)$$

But $T = T_s$ is a constant. Therefore, p/ρ is also a constant. Hence,

$$\frac{p_1}{p_2} = \frac{\rho_1}{\rho_2}$$

Thus, the variation of pressure and density in the stratosphere obeys the law:

所以，平流层中压力和密度变化遵循如下规律：

$$\frac{p_2}{p_1} = \frac{\rho_2}{\rho_1} = \exp\left[\frac{g(h_1 - h_2)}{RT_s}\right] \tag{3.22}$$

3.7.3 Calculations on the Troposphere

3.7.3 对流层计算

It is well-known that the variation of atmospheric temperature with altitude in troposphere is given by

众所周知，对流层中大气温度随高度的变化可表示为

$$T = T_0 - \lambda h \tag{3.23}$$

where T_0 is the absolute temperature at mean sea level and λ is the "temperature lapse rate", the rate of decrease of temperature with altitude in Kelvin per meter. For earth's atmosphere the lapse rate is 6.5 K per km

式中，T_0 是基准海平面处的绝对温度；λ 是"温度递减率"，即温度随高度降低率，单位为 K/m。对于地球大气，温度递减率是 6.5K/km。

From Equations (3.16) and (1.2) we have

由式（3.16）和式（1.2）得

$$\frac{\mathrm{d}p}{p} = -\frac{g}{RT}\mathrm{d}h$$

Substituting Equation (3.23) into this, we get

把式（3.23）代入上式，得

$$\frac{\mathrm{d}p}{p} = -\frac{g}{R}\frac{\mathrm{d}h}{T_0 - \lambda h}$$

Integrating this between altitudes h_1 and h_2 and the corresponding pressure p_1 and p_2, we obtain

在高度 h_1 和 h_2 以及相应的 p_1 和 p_2 之间对上式积分，得到

$$[\ln p]_{p_1}^{p_2} = -\frac{g}{\lambda R}[\ln(T_0 - \lambda h)]_{h_1}^{h_2} \tag{3.24}$$

Now, since $T_0 - \lambda h = T$, Equation (3.24) may be expressed as

现在，由于 $T_0 - \lambda h = T$，所以式（3.24）可表示为

$$\ln\left(\frac{p_2}{p_1}\right) = \frac{g}{\lambda R}\ln\left(\frac{T_2}{T_1}\right)$$

or

或

$$\frac{p_2}{p_1} = \left(\frac{T_2}{T_1}\right)^{\frac{g}{\lambda R}} \tag{3.25}$$

This gives the variation of pressure with temperature and hence altitude. From the equation of state, we have

上式给出了压力随温度，当然也是随高度的变化关系。由状态方程，有

$$\frac{p_2}{p_1} = \frac{T_2}{T_1}\frac{\rho_2}{\rho_1}$$

| Therefore, | 因此， |

$$\frac{\rho_2}{\rho_1} = \frac{p_2}{p_1} \cdot \frac{T_1}{T_2} = \left(\frac{T_2}{T_1}\right)^{g/\lambda R} \frac{T_1}{T_2} = \left(\frac{T_2}{T_1}\right)^{(g/\lambda R)-1}$$ (3.26)

$$\frac{\rho_2}{\rho_1} = \left(\frac{T_2}{T_1}\right)^{(g/\lambda R)-1} = \left(\frac{T_2}{T_1}\right)^{(g-\lambda R)/\lambda R}$$

| From Equations (3.25) and (3.26), it follows that | 由式（3.25）和式（3.26），得出 |

$$\frac{p_2}{p_1} = \left(\frac{\rho_2}{\rho_1}\right)^{g/(g-\lambda R)}$$ (3.27)

| The conditions represented by subscripts 1 and 2 are selected arbitrarily, and hence are quite general. Therefore, Equation (3.27) may be written as | 下标1和下标2代表的条件是任意选取的，因此具有普遍性。于是，式（3.27）可以写为 |

$$p = k\rho^{g/(g-\lambda R)}$$ (3.28)

| where k is a constant. | 式中，k 是常数。 |
| For the International Standard Atmosphere (ISA) the value of lapse rate is 0.0065K/m. With this value for λ and the value of 287.26J/(kg·K) for gas constant R, the Equations (3.25), (3.26) and (3.28) became | 对于国际标准大气（ISA），递减率 λ 的值是 0.0065K/m。将这一 λ 值和气体常数 R 的 287.26J/(kg·K) 值代入，式（3.25）、式（3.26）和式（3.28）变为 |

$$\frac{p_2}{p_1} = \left(\frac{T_2}{T_1}\right)^{5.254}$$ (3.29)

$$\frac{\rho_2}{\rho_1} = \left(\frac{T_2}{T_1}\right)^{4.254}$$ (3.30)

| and | 和 |

$$p = k\rho^{1.235}$$ (3.31)

| Note: It should be noted that, in the above discussion the gravitational acceleration is considered to be constant. Actually, it varies inversely with the square of the distance from the centre of earth. This correction appears to be insignificant (less than 0.5 percent) within the stratosphere. | 注意：上述讨论中重力加速度被认为是常数，实际上，它与到地心距离的平方成反比，只不过在平流层中这一修正是微乎其微的（小于0.5%）。 |
| **Example 3.1** Calculate the pressure and density at altitudes of | **例 3.1** 在 ISA 中计算 5km 和 22km |

5km and 22km in the ISA. The sea level pressure and temperature are 101325Pa and 288K, and temperature lapse rate is 6.5K/km. Assume that the tropopause is at 11km and the temperature here is 216.5K.

Solution

(a) At 5km altitude:

$$T = T_0 - \lambda h = 288 - 6.5 \times 5 = 255.5\text{K}$$

Therefore,

$$\frac{101325}{p_5} = \left(\frac{288}{255.5}\right)^{5.254} = 1.876$$

Therefore,

$$p_5 = \frac{101325}{1.876} = 54011.2\text{Pa}$$

Also, from the equation of state

$$\rho_5 = \frac{p_5}{RT_s} = \frac{54011.2}{287.26 \times 255.5} = 0.736\text{kg}/\text{m}^3$$

(b) At 22km altitude:

At this altitude the change in characteristics between the temperature and the stratosphere must be taken into account. Therefore, it is necessary to calculate the conditions at the troposphere using the troposphere equations and then proceed using the equations for the stratosphere.

Let us use subscript "11" to denote values at the troposphere of 11km. In the tropopause, temperature $T_{11} = 216.5\text{K}$. Therefore,

$$\frac{101325}{p_{11}} = \left(\frac{288}{216.5}\right)^{5.254} = 4.48$$

Thus,

$$p_{11} = \frac{101325}{4.48} = 22617\text{Pa}$$

In the stratosphere

$$\frac{p_{11}}{p_{22}} = \exp\left[\frac{(22000-11000) \times 9.81}{287.26 \times 216.5}\right] = \exp(1.735) = 5.669$$

Therefore,

（b）在 22km 高度：

在这一高度，必须考虑温度和平流层之间特性的变化，因此有必要用对流层方程计算对流层的边界条件，然后继续用这一方程对平流层进行计算。

用下标 11 表示对流层 11km 处的值。在对流层顶，温度 T_{11} = 216.5K。所以，

于是，

在平流层

因此，

$$p_{22} = \frac{22617}{5.669} = 3990 \text{Pa}$$

| From state equation | 由状态方程，有 |

$$\rho_{22} = \frac{p_{22}}{RT_{22}} = \frac{3990}{287.26 \times 216.5} = 0.0642 \text{kg/m}^3$$

Example 3.2

A compressed air tank is fitted with a window of 200mm diameter. A U-tube manometer filled with mercury, connected between the tank and atmosphere, reads 2.2m. Calculate the total load acting on the bolts securing the window.

Solution

The pressure in the tank is

例 3.2

一个压缩空气罐上装有一个直径200mm的窗口，一个U形管汞柱压力计与空气罐和大气相连，读数为2.2m，计算作用在窗口紧固螺栓上的总负载力。

解

空气罐中压力为

$$p_{\text{Hg}} = 2.2\text{m}（\text{gauge pressure} 表压力）$$

One atmosphere is ISA in 760mmHg, which is equal to 101325Pa. Therefore, the pressure in the tank becomes

ISA中一个大气压是760mmHg，等于101325Pa。因此，罐中压力为

$$p = \frac{2.2}{0.76} \times 101325 = 293309\text{Pa}（\text{gauge pressure} 表压力）$$

| Force due to this pressure acting on the window is | 由压力产生的作用在窗口上的力为 |

$$F = p \times A$$

| where A is the surface area of the window. Thus, | 式中，A是窗口表面面积。所以， |

$$F = 293309 \times \frac{\pi \times 0.2^2}{4} = 9214.6\text{N}$$

This is the total load acting on the bolts securing the window.

Note: In this problem, the gauge pressure is used to calculate the force and not the absolute pressure. This is because atmospheric pressure is acting on the outside surface of the window. Hence, the pressure due to atmosphere at the inner and outer surfaces of the window gets canceled.

Example 3.3

Assume the atmospheric air to be an ideal gas of

这就是作用在窗口紧固螺栓上的总负载力。

注意：这一问题中，使用表压力而不是绝对压力来计算力，因为有大气压作用在窗口外部，所以大气压产生的作用力在窗口外部和内部相互抵消。

例 3.3

假设大气是具有恒定比热

constant specific heat ratio $\gamma = c_p/c_v$. Also assume the gravitational acceleration to be constant over the range of the atmosphere. Let $z = 0$ at sea level, and T_0, p_0, ρ_0 to be the absolute temperature, pressure, and density of atmospheric air, respectively, at $z = 0$.
(a) Assuming that, the thermodynamic variables of the atmosphere are related in the same way they would be for an adiabatic process, find $p(z)$ and $\rho(z)$.
(b) Show that for this case no atmosphere exists above a z_{max}, given by, $z_{max} = \dfrac{\gamma}{\gamma-1}\dfrac{RT_0}{g}$, where R is the gas constant for air.

Solution

(a) By Equation (3.19), we have

$$\frac{\mathrm{d}p}{p} = -\frac{g\mathrm{d}z}{RT}$$

that is

$$\frac{\mathrm{d}p}{\mathrm{d}z} = -g\rho(z) \qquad (3.32)$$

since $p/RT = \rho$ by the equation of state.

By adiabatic process relation, we have

$$\frac{p}{\rho^\gamma} = \frac{p_0}{\rho_0^\gamma}$$

or

$$\mathrm{d}p = \gamma\rho(z)^{\gamma-1}\mathrm{d}\rho(z)\frac{p_0}{\rho_0^\gamma}$$

Replacing $\mathrm{d}p$ with the expression in Equation (3.32), we obtain

$$-g\rho(z)\mathrm{d}z = \gamma\rho(z)^{\gamma-1}\mathrm{d}\rho(z)\frac{p_0}{\rho_0^\gamma}$$

$$\rho(z)^{\gamma-2}\mathrm{d}\rho(z) = -\frac{g\rho_0^\gamma}{\gamma p_0}\mathrm{d}z$$

By state equation $p = \rho RT$, thus, we have

$$\rho(z)^{\gamma-2}\mathrm{d}\rho(z) = -\frac{g\rho_0^\gamma}{\gamma p_0 RT_0}\mathrm{d}z$$

$$\rho(z)^{\gamma-2}\,\mathrm{d}\rho(z) = -\frac{g\rho_0^{\gamma-1}}{\gamma RT_0}\mathrm{d}z$$

Integrating between $z = 0$ and z, we obtain	在 0 到 z 的区间积分，可得

$$\left[\frac{\rho^{\gamma-1}}{\gamma-1}\right]_0^z = -\rho_0^{\gamma-1}\frac{gz}{\gamma RT_0}$$

$$\rho(z)^{\gamma-1} - \rho(0)^{\gamma-1} = \rho(z)^{\gamma-1} - \rho_0^{\gamma-1}$$

$$\rho(z)^{\gamma-1} - \rho_0^{\gamma-1} = -\rho_0^{\gamma-1}\frac{\gamma-1}{\gamma}\frac{gz}{RT_0}$$

$$\rho(z)^{\gamma-1} = \rho_0^{\gamma-1}\left(1 - \frac{\gamma-1}{\gamma}\frac{gz}{RT_0}\right)$$

$$\rho(z) = \rho_0\left(1 - \frac{\gamma-1}{\gamma}\frac{gz}{RT_0}\right)^{1/(\gamma-1)} \quad (3.33)$$

Also,	又有

$$\frac{p}{\rho^\gamma} = \frac{p_0}{\rho_0^\gamma}$$

$$\frac{\rho}{\rho_0} = \left(\frac{p}{p_0}\right)^{1/\gamma}$$

Therefore, Equation (3.33) becomes	因此，式（3.33）变为

$$\left(\frac{p}{p_0}\right)^{1/\gamma} = \left(1 - \frac{\gamma-1}{\gamma}\frac{gz}{RT_0}\right)^{1/(\gamma-1)}$$

$$p(z) = p_0\left(1 - \frac{\gamma-1}{\gamma}\frac{gz}{RT_0}\right)^{\gamma/(\gamma-1)}$$

(b) At $z_{\max}$, $\rho = 0$, thus, from Equation (3.33), we have	（b）在 $z_{\max}$ 处，$\rho = 0$，根据式（3.33），有

$$z_{\max} = \frac{\gamma}{\gamma-1}\frac{RT_0}{g}$$

3.8 Hydrostatic Force on Submerged Surfaces

3.8 浸没表面上的静压力

Consider a two-dimensional curved plate, submerged in a stationary incompressible fluid, as shown in Figure 3.5. Examine the upper surface of the plate. The pressure force acting on the elemental area

取一个浸没在静止不可压缩流体中的二维曲面，如图 3.5 所示，观察曲面的上表面。作用在微元面积 $\mathrm{d}A$ 上的压力为 $-p\mathrm{d}A$，

dA is $-p$ dA, that is, inwards. By Equation (3.17), the pressure p is	即压力作用于曲面内表面，由式（3.17），压力p为
$$p = p_{atm} + \rho g h$$	
In terms of gauge pressure,	用表压力表示为
$$p_g = \rho g h$$	
Therefore, the force acting on the elemental area dA is	所以，作用在微元面积 dA 上的静压力为
$$dF = -\rho g h dA \qquad (3.34)$$	

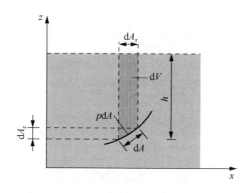

Figure 3.5　Force on a submerged body

图 3.5　作用在浸没体上的力

The total force acting on the upper surface of the plate, F, is obtained by integrating dF over the area as	通过对整个面积上的dF积分，求得作用在曲面上表面的合力 F 为
$$F = -\rho g \iint_A h dA \qquad (3.35)$$	
Equation (3.35) gives the resultant force acting at the upper surface of the plate due to the gauge pressure. The vertical component of the resultant force is given by	式（3.35）给出的是由表压力计算得到的上表面作用力，作用力的垂直分量可写为
$$F_z = -\rho g \iint_A h(k \cdot dA) = -\rho g \iint_A h dA_x \qquad (3.36)$$	
where k is the unit vector, along the vertical direction, and dA_x is the projected area in vertical direction. But (hdA_x) is the volume dV of fluid prism that stands vertically on the area element dA. Therefore, the vertical force on the upper surface is given by	式中，k 是沿垂直方向的单位矢量；dA_x 是竖直方向上的投影面积。(hdA_x)是面积微元dA上竖直方向的流体柱体积dV。因此，上表面受到的竖直方向作用力为

$$F_z = -\rho g V \qquad (3.37)$$

The negative sign in Equation (3.37) shows that F_z acts downwards. For this plate, the vertical force on the bottom surface is also the same as F_z given by Equation (3.37), except that it acts upwards. Similarly, the horizontal component of the force is given by

式(3.37)中负号表示F_z作用方向向下。对于这一曲面，下表面上的垂直作用力除了作用方向向上外，也和F_z一样由式(3.37)给出。同理，力的水平分量可表示为

$$F_x = -\rho g \iint_A h(i \cdot \mathrm{d}A) = -\rho g \iint_A h \mathrm{d}A_z \qquad (3.38)$$

where i is the unit vector along the x-direction and $\mathrm{d}A_z$ is the area projected in the x-direction. Thus, the horizontal forces on the curved surfaces are simply estimated with the projected area of the surfaces in the respective direction.

式中，i 是沿 x 方向的单位矢量；$\mathrm{d}A_z$ 是 x 方向投影面积。因此，曲面上水平方向的作用力用相应方向上的投影面积计算即可。

3.9 Buoyancy

3.9 浮力

The buoyant force on a body is defined as the vertical force acting on it due to the fluid or fluids in contact with the body. A body in flotation is in contact only with fluids, and the surface force from the fluids is in equilibrium with the force of gravity on the body.

作用在物体上的浮力可定义为流体或和物体接触的流体作用在物体上的竖直方向的作用力。漂浮的物体只和流体接触，则由流体产生的表面力和作用在物体上的重力是相平衡的。

Consider a three-dimensional body, completely submerged in a fluid of density ρ, as shown in Figure 3.6.

取一个三维物体，完全浸没在密度为 ρ 的流体中，如图 3.6 所示。

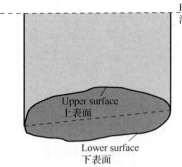

Figure 3.6　A body experiencing buoyancy

图 3.6　受浮力作用的物体

By Equation (3.37), the downward force, due to fluid acting on the upper surface of the body is $\rho g V_1$, where V_1 is the volume of fluid prism that stands on the upper surface and extends to the fluid level. Similarly, the upward force due to the fluid acting on the lower surface is $\rho g V_2$, where V_2 is the sum of the the volumes V_1 and the volume of the body. The net upward force, due to fluid, acting on the body, termed as buoyancy, is given by

$$F_B = \rho g (V_2 - V_1) \qquad (3.39)$$

where $(V_2 - V_1)$ is the volume of the body V_s, and therefore, the buoyant force becomes

$$F_B = \rho g V_s \qquad (3.40)$$

Equation (3.40) is the mathematical form of the law of buoyancy found by Archimedes which states that, "*a body immersed in a fluid experiences a buoyant force equal to the weight of the fluid displaced by it*". From Equation (3.40) it is also seen that, when ρ is constant, the buoyancy force does not depend on the depth of submergence.

3.10 Summary

When all elements in a fluid are at rest with respect to a reference coordinate system, the fluid is considered to be static. In this chapter, we mainly talked about the basic equations for fluid statics. Before that, some basic concepts, like the quantities of scalar, vector and tensor, body force and surface force were introduced. Some examples to solve the static problems of fluids were also given for practice in this chapter.

3.11 Exercises

Problem 3.1 What is the pressure difference between the points *A* and *B* in the tanks shown in Figure 3.7?

[Ans: $p_A - p_B = \rho_{Hg}gh_1 - (h_2 + h_3)\rho_{H_2O}g$]

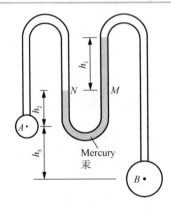

Figure 3.7　Two bulbs connected by a mano meter

Problem 3.2 Determine the pressure at a depth of 10km below the surface of a sea. Take the average specific gravity of water to be 1.3.

[Ans: 127.63MPa]

Problem 3.3 Determine the pressure at point *A* in the tank, shown in Figure 3.8.

[Ans: 119.5716kPa]

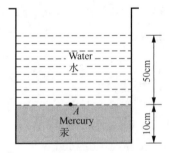

Figure 3.8　A tank with mercury and water

Problem 3.4 Consider a U-tube and a funnel arrangement, shown in Figure 3.9. Mercury is poured into the funnel to trap the air in the tube. The tube has 10mm inside diameter and 1m total length. Assuming that the trapped air is compressed isothermally, determine h at which the funnel will begin to overflow.

[Ans: 0.3942m]

题 3.4 取一个 U 形管和一个漏斗组合如图 3.9 所示。在漏斗中注入汞封住管中空气，管内径 10mm，总长 1m。假设封住的气体被绝热压缩，求漏斗开始溢出时的高度 h。

【答：0.3942m】

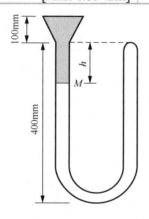

Figure 3.9 A U-tube and funnel arrangement

图 3.9 一个 U 形管和漏斗组合

Problem 3.5 Find the total force on door AB, shown in Figure 3.10, and the moment of this force about the bottom of the door. The width of the door is 2m.

[Ans: 2601612N, 2601612N·m]

题 3.5 计算图 3.10 所示门 AB 上的总作用力以及绕门底部的力矩，门宽度为 2m。

【答：2601612N，2601612N·m】

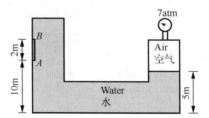

Figure 3.10 A reservoir with a door

图 3.10 带门的水箱

Problem 3.6 A nitrogen tank contains 8kg of nitrogen at 10MPa (absolute) and 30℃. If the tank is spherical in shape, determine its volume and diameter.

[Ans: 0.072m³, 0.516m]

题 3.6 一个氮气罐装有 10MPa（绝对压力）、30℃的氮气 8kg，如果罐是球形的，计算它的体积和直径。

【答：0.072m³，0.516m】

Problem 3.7 A swimmer can move at a maximum steady speed of v_r = 2m/s in still water. To do so, the swimmer must produce enough power to overcome water resistance. The drag force acting on the swimmer is estimated to be $F_D = kv^2$, where, $k = 25\text{N}\cdot\text{s}^2/\text{m}^2$, and v is the speed of the swimmer, relative to the water. (a) Evaluate the power produced by the swimmer in still water. If the swimmer can maintain the same power output in a river current that moves at 3.5km/h, estimate the maximum speed the swimmer can reach, while (b) swimming upstream and (c) swimming downstream.

[Ans: (a) 200W; (b) 1.03m/s; (c) 2.97m/s]

Problem 3.8 If a Pitot-static tube connected to a water manometer has to measure an air speed of 0.6m/s, within ±1 percent accuracy, under standard sea level conditions, what should be the sensitivity of the manometer?

[Ans: 0.022mm of water]

Problem 3.9 A pressure gauge connected to a tank reads 500kPa, at a place where the atmospheric pressure is 100kPa. Determine the tank pressure.

[Ans: 600kPa]

Problem 3.10 A vacuum gauge connected to a tank reads 50kPa at a place where the barometric reading is 760mmHg. Determine the absolute pressure in the tank. Take $\rho_{Hg} = 13.6\text{kg/m}^3$.

[Ans: 51.396kPa]

Problem 3.11 A 10m diameter spherical balloon is filled with helium gas of density 0.3kg/m³. If the balloon has to carry a load of 200kg, determine the acceleration of the balloon when it is first released and the maximum amount of load that the balloon can carry.

题 3.7 一个游泳者可以在静止的水中以 v_r=2m/s 的最大稳定速度移动，为此，游泳者必须产生足够的能量来克服水的阻力，游泳者所受到的阻力可估算为 $F_D=kv^2$，其中 $k=25\text{N}\cdot\text{s}^2/\text{m}^2$，$v$ 是相对于水的游泳速度。（a）计算游泳者在静止水中产生的动力。如果游泳者用同样的动力在以 3.5km/h 流速流动的河流中移动，计算（b）逆流和（c）顺流时游泳者的最大速度。

【答：（a）200W；（b）1.03m/s；（c）2.97m/s】

题 3.8 如果一个和水压表连接的皮托管必须要测量 0.6m/s 的空气速度，在精度范围为 ±1%，标准海平面条件下，压力表的灵敏度应该是多少？

【答：0.022mm H₂O】

题 3.9 容器上连接一个压力表，读数为 500kPa，位于 100kPa 大气压处，计算容器中的压力值。

【答：600kPa】

题 3.10 容器上连接一个真空压力表，读数为 50kPa，位于压力计读数为 760mmHg 处，计算容器中绝对压力，取 ρ_{Hg}=13.6kg/m³。

【答：51.396kPa】

题 3.11 一个直径 10m 的圆气球充满密度为0.3kg/m³ 的氢气，如果气球要携带 200kg 负载，计算气球被释放时的加速度以及气球能够携带的最大负载。假设

Assume the air density to be 1.225kg/m³, and neglect the weight of the ropes and the cage.

[Ans: 7.81m/s²]

Problem 3.12 Determine the pressure that can act on a diver at 50m depth in a sea. Assume the specific gravity of seawater to be 1.03 and the atmospheric pressure to be 101kPa.

[Ans: 606.215kPa]

Problem 3.13 A perfect gas at constant temperature is at rest in equilibrium in a uniform gravitational field. Find the pressure as a function of height z, given that, at $z = 0$, $p = p_0$.

[Ans: $p_0 \exp(-gz/RT)$]

Problem 3.14 A mercury manometer connected to a wall pressure tap of a duct with air flow shows a pressure of 25mmHg suction. If the atmospheric pressure is 101kPa, determine the absolute static pressure at the duct wall.

[Ans: 97.665kPa]

Problem 3.15 A cylinder contains a fluid at a gauge pressure of 350kPa. What would be the absolute pressure in the cylinder if the atmospheric pressure is 101.3kPa? Express the pressure in terms of (a) water and (b) mercury.

[Ans: 451.3kPa; (a) 35.68m; (b) 2.62m]

Problem 3.16 A spherical balloon of diameter 1.5m and weight 8.50kN is anchored to the sea floor with a cable. The balloon is completely immersed in the water. If the specific weight of seawater is 10.1kN/m³, calculate the tension of the cable.

[Ans: 9.35kN]

题 3.12 计算海面下 50m 处潜水员受到的压力。假设海水的比重是 1.03，大气压力为 101kPa。

【答：606.215kPa】

题 3.13 恒温下的完全气体在均匀重力场中处于静止平衡状态，给出压力与高度 z 的函数关系，设 $z=0$ 处，$p=p_0$。

【答：$p_0 \exp(-gz/RT)$】

题 3.14 与气流流过的管道壁面测压口相连的压力计显示 25mmHg 吸力，如果大气压是 101kPa，计算管壁上的绝对静压。

【答：97.665kPa】

题 3.15 缸筒中装有表压力为 350kPa 的流体，如果大气压力为 101.3kPa，缸筒中绝对压力是多少？分别用（a）水柱和（b）汞柱表示压力。

【答：451.3kPa；（a）35.68m；（b）2.62m】

题 3.16 一个直径 1.5m、重 8.50kN 的圆气球通过缆绳拴在海底，气球完全浸没在水中，如果海水比重为 10.1kN/m³，计算缆绳拉力。

【答：9.35kN】

References
参考文献

[1] Rathakrishnan E . Instrumentation, Measurements, and Experiments in Fluids[M]. 2nd. Boca Raton: CRC Press, 2017.

Chapter 4   Kinematics and Dynamics of Fluid Flow

第 4 章   流体运动学和动力学

4.1 Introduction

Kinematics of fluid flow deals with the motion of fluids in terms of displacements, velocities and accelerations without reference to the forces that cause the motion. While fluid dynamics deals with the effects of fluid and its surroundings on the motion of fluid with reference to the interaction forces. This chapter introduces the description of fluid flow, including Lagrangian and Eulerian methods, as well as graphical descriptions of fluid motion. Basic and subsidiary equations for fluid flow are summarized in this chapter. The rotational and irrotational motion of fluid are briefly discussed. A combination of simple flows is analyzed with the dynamics theory of fluid flow.

4.2 Description of Fluid Flow

4.2.1 Lagrangian and Eulerian Methods

Basically two treatments are followed for fluid flow analysis. They are the Lagrangian and Eulerian descriptions. Lagrangian method describes the motion of each particle of the flow field in a separate and discrete manner. For example, the velocity of the nth particle of an aggregate of particles, moving in space, can be specified by the scalar equations:

$$(V_x)_n = f_n(t) \tag{4.1a}$$
$$(V_y)_n = g_n(t) \tag{4.1b}$$
$$(V_z)_n = h_n(t) \tag{4.1c}$$

where V_x, V_y, V_z are the velocity components in

4.1 引言

流体运动学以位移、速度和加速度形式求解流体运动，不涉及导致流体运动的作用力。而流体动力学求解流体与其周围介质之间相互的作用力对流体流动的影响。本章介绍流体流动的描述方法，包括拉格朗日法、欧拉法和图形化方法，总结流体流动的基本方程和补充方程，简要讨论流体的有旋运动和无旋运动，并用流体动力学理论分析一个简单流动的组合。

4.2 流体流动的描述

4.2.1 拉格朗日法和欧拉法

流体流动分析基本遵循两种方法，它们是拉格朗日描述和欧拉描述。拉格朗日法用一种分散和离散的方式描述流场中每个质点的运动。例如，质点集合中第 n 个质点在空间中运动的速度可通过标量方程组给定：

(4.1a)
(4.1b)
(4.1c)

式中，V_x、V_y、V_z 分别是 x、y、

x-, y-, z-directions, respectively. They are independent of the space coordinates, and are functions of time only. Usually, the particles are denoted by the space point they occupy at some initial time t_0. Thus, $T(x_0, t)$ refers to the temperature at time t of a particle which was at location x_0 at time t_0.

This approach of identifying material points, and following them along is also termed the particle or material description. This approach is usually preferred in the description of low-density flow fields (also called rarefied flows) and in describing the motion of moving solids, such as a projectile and so on. However, for a deformable system like a continuum fluid, there are infinite number of fluid elements whose motion has to be described, the Lagrangian approach becomes unmanageable. For such cases, we can employ spatial coordinates to help to identify particles in a flow. The velocity of all particles in a flow field, therefore, can be expressed in the following manner:

z方向上的速度分量。它们与空间坐标无关，仅是时间的函数。通常，用质点在某一初始时刻 t_0 所在的空间位置来标注质点。因此，$T(x_0, t)$ 指在 t_0 时刻位于 x_0 位置的一个质点在 t 时刻的温度。

这种标记质点且这个标记一直跟随着每个质点的方法也被称为质点描述或物质描述。这种方法更适用于低密度流场（也称为稀薄流）的描述，以及描述运动着的固体，例如弹头等。然而，对于连续流体这样一个可变形系统，有无穷多个需要描述其运动的流体微元，拉格朗日法就变得难以处理了。对于这样的情况，我们可以利用空间坐标来帮助区分流场中的质点。因此，可以用如下的方式来表示流场中所有质点的速度：

$$V_x = f(x, y, z, t) \quad (4.2a)$$
$$V_y = g(x, y, z, t) \quad (4.2b)$$
$$V_z = h(x, y, z, t) \quad (4.2c)$$

This is called the Eulerian or field approach. If properties and flow characteristics at each position in space remain invariant with time, the flow is called steady flow. A time dependent flow is referred to as unsteady flow. The steady flow velocity field would then be given as

这种描述方法称为欧拉法或流场法。如果空间中每一点的性质和流动特征是不随时间变化的，这种流动就称为定常流动。随时间变化的流动是非定常流动。定常流动的速度场可以表示为

$$V_x = f(x, y, z) \quad (4.3a)$$
$$V_y = g(x, y, z) \quad (4.3b)$$
$$V_z = h(x, y, z) \quad (4.3c)$$

4.2.2　Local and Material Rates of Change

4.2.2　当地导数和随体（物质）导数

The rate of change of a property measured by probes at fixed locations are referred to as local rates

在一个固定位置上探头测得的一种性质的变化率是当地

of change, and the rate of change of properties experienced by a material particle is termed the material or substantive rates of change.

The local rate of change of a property η is denoted by $\partial \eta(x,t)/\partial t$, where it is understood that x is held constant. The material rate of change of property η shall be denoted by $D\eta/Dt$. If η is the velocity V, then DV/Dt is the rate of change of velocity for a fluid particle and thus, is the acceleration that the fluid particle experiences. On the other hand, $\partial V/\partial t$ is just a local rate of change of velocity recorded by a stationary probe. In other words, DV/Dt is the particle or material acceleration and $\partial V/\partial t$ is the local acceleration.

For a fluid flowing with an uniform velocity V_∞, it is possible to write the relation between the local and material rates of change of property η as

$$\frac{\partial \eta}{\partial t} = \frac{D\eta}{Dt} - V_\infty \frac{\partial \eta}{\partial x} \tag{4.4}$$

Thus, the local rate of change of η is due to the following two effects:

(1) The change of property of each particle with time.

(2) The combined effect of the spatial gradient of that property and the motion of the fluid.

When a spatial gradient exists, the fluid motion brings different particles with different values of η to the probe, thereby modifying the rate of change sensed by the probe. This effect is termed convection effect. Therefore, $V_\infty(\partial \eta/\partial x)$ is referred to as the convective rate of change of η. Even though Equation (4.4) has been obtained with uniform velocity V_∞, note that, in the limit $\delta t \to 0$ it is only the local velocity V which enters the analysis and Equation (4.4) becomes

$$\frac{\partial \eta}{\partial t} = \frac{D\eta}{Dt} - V \frac{\partial \eta}{\partial x} \tag{4.5}$$

Equation (4.5) can be generalized for a three-dimensional space as

导数，一个质点所经历的性质变化率称为随体导数或物质导数。

一个性质 η 的当地导数由 $\partial \eta(x,t)/\partial t$ 表示，其中 x 是保持不变的。性质 η 的随体导数表示为 $D\eta/Dt$。如果 η 是速度 V，则 DV/Dt 是一个流体质点速度的导数，即流体质点所受的加速度。另一方面，$\partial V/\partial t$ 只是通过静止的探头测得的当地速度的导数。换句话说，DV/Dt 是质点或随体加速度，而 $\partial V/\partial t$ 是当地加速度。

对于一个以速度 V_∞ 匀速流动的流体，可以写出性质 η 的当地导数和随体导数之间的关系式

因此，η 的当地导数归因于如下两方面的影响：

（1）每个质点性质随时间的变化。

（2）某个性质的空间梯度和流体运动共同的作用。

当存在空间梯度时，流体运动使探头接触到具有不同 η 值的不同质点，所以，改变了探头所检测 η 值的变化率，这个效应称为迁移效应。因此，$V_\infty(\partial \eta/\partial x)$ 称为 η 的迁移导数。尽管式（4.4）是在 V_∞ 匀速的情况下获得的，注意，在 $\delta t \to 0$ 极限情况下，分析中仅可代入当地速度 V，因此式（4.4）变为

式（4.5）可被推广到三维空间中，有

$$\frac{\partial \eta}{\partial t} = \frac{D\eta}{Dt} - (V \cdot \nabla)\eta \qquad (4.6)$$

where ∇ is the gradient operator $\nabla \equiv i\partial/\partial x + j\partial/\partial y + k\partial/\partial z$ and $(V \cdot \nabla)$ is a scalar product, $(V \cdot \nabla) = V_x \partial/\partial x + V_y \partial/\partial y + V_z \partial/\partial z$. Equation (4.6) is usually written as	式中，∇ 是梯度算子，$\nabla \equiv i\partial/\partial x + j\partial/\partial y + k\partial/\partial z$；$(V \cdot \nabla)$ 是一个标量积，$(V \cdot \nabla) = V_x \partial/\partial x + V_y \partial/\partial y + V_z \partial/\partial z$。式（4.6）通常写为

$$\frac{D\eta}{Dt} = \frac{\partial \eta}{\partial t} + (V \cdot \nabla)\eta \qquad (4.7)$$

when η is the velocity of a fluid particle, DV/Dt gives acceleration of the fluid particle and the resultant equation is	当 η 是流体质点的速度时，DV/Dt 为流体质点的加速度，则得到方程

$$\frac{DV}{Dt} = \frac{\partial V}{\partial t} + (V \cdot \nabla)V \qquad (4.8)$$

Equation (4.8) is known as Euler's acceleration formula.

Note that, the Euler's acceleration formula is essentially the link between the Lagrangian and Eulerian descriptions of fluid flow.

4.2.3 Graphical Description of Fluid Motion

The four important concepts for visualizing or describing flow fields are the concept of pathline, the concept of streakline, the concept of streamline and the concept of timeline.

1. Pathline

Pathline may be defined as a line in the flow field describing the trajectory of a given fluid particle. From the Lagrangian view point, namely, for a closed system with a fixed identifiable quantity of mass, the independent variables are the initial position, with which each particle is identified, and the time. Hence, the locus of the same particle over a time period from t_0 to t_n is called the pathline.

2. Streakline

Streakline may be defined as the instantaneous

式（4.8）就是欧拉加速度方程。

注意：欧拉加速度方程本质上是流体流动的拉格朗日描述和欧拉描述之间的纽带。

4.2.3 流体运动的图形化描述

流场可视化和描述流场的四个重要概念是迹线、脉线、流线和时间线。

1. 迹线

迹线可以定义为流场中一条描述某一给定流体质点运动轨迹的线。从拉格朗日法的观点出发，即对于一个具有一定可确定质量的封闭系统，独立变量是用于标记每个质点的初始位置和时间的。因此，同一个质点从 t_0 到 t_n 时间段的轨迹称为迹线。

2. 脉线

脉线被定义为已流过流入点

loci of all the fluid elements that have passed the point of injection at some earlier time. Consider a continuous tracer injection at a fixed point Q in space. The connection of all elements passing through the point Q over a period of time is called the streakline.

3. Streamlines

Streamlines are imaginary lines, in a fluid flow, drawn in such a manner that the flow velocity is always tangential to it. Flows are usually depicted graphically with the aid of streamlines. These are imaginary lines in the flow field such that the velocity at all points on these lines are always tangential. Streamlines proceeding through the periphery of an infinitesimal area at some instant of time t will form a tube called streamtube, which is useful in the study of fluid flow.

From Eulerian view point, namely, for an open system with constant control volume, all flow properties are functions of a fixed point in space and time, if the process is transient. The flow direction of various particles at time t_i forms streamline. The pathline, streamline and streakline are different in general but coincide in a steady flow.

4. Timelines

In modern fluid flow analysis, yet another graphical representation, namely timeline is used. When a pulse input is periodically imposed on a line of tracer source placed normal to a flow, a change in the flow profile can be observed. The tracer image is generally termed timeline. Timelines are often generated in the flow field to aid the understanding of flow behavior such as the velocity and velocity gradient.

From the above-mentioned graphical descriptions, We can infer that,

(1) There can be no flow through the lateral surface of the streamtube.

(2) An infinite number of adjacent streamlines arranged to form a finite crosssection is often called a bundle of streamlines.

(3) Streamtube is an Eulerian (or field) concept.

(4) Pathline is a Lagrangian (or particle) concept.

(5) For steady flows, the pathline, streamline and streakline are identical.

4.3 Basic and Subsidiary Laws

In the range of engineering interest, four basic laws must be satisfied for any continuous medium. They are:

(1) Conservation of matter (continuity equation).

(2) Newton's second law (momentum equation).

(3) Conservation of energy (first law of thermodynamics).

(4) Increase of entropy principle (second law of thermodynamics).

In addition to these primary laws, there are numerous subsidiary laws, sometimes called constitutive relations, which apply to specific types of media or flow processes (for example, equation of state for perfect gas, Newton's viscosity law for certain viscous fluids, isentropic and adiabatic process relations are some of the commonly-used subsidiary equations in flow physics).

4.3.1 System and Control Volume

In employing the basic and subsidiary laws, any

可以推断出：

（1）流动不会穿过流管的侧表面。

（2）通常把由无限多的相邻流线组成的一个有限横截面称为流束。

（3）流管是欧拉法（或场）概念。

（4）迹线是拉格朗日法（或质点）概念。

（5）对于定常流动，迹线、流线和脉线是同一条线。

4.3 基本定律和补充定律

在工程实际问题中，任何连续介质都必然满足4个基本定律。它们是：

（1）物质守恒（连续性方程）。

（2）牛顿第二定律（动量方程）。

（3）能量守恒（热力学第一定律）。

（4）熵增加原理（热力学第二定律）。

除了这些基本定律外，还有很多适用于特殊种类物质或流动过程，有时称为本构关系的补充定律（例如，完全气体的状态方程、某些黏性流体的牛顿黏性定律、等熵和绝热过程关系都是流体力学中一些常用的补充方程）。

4.3.1 系统和控制体

在应用基本定律和补充定律

one of the following modes of application may be adopted.

(1) The activities of each and every given element of mass must be such that it satisfies the basic laws and the pertinent subsidiary laws.

(2) The activities of each and every elemental volume in space must be such that the basic laws and the pertinent subsidiary laws are satisfied.

In the first case, the laws are applied to an identified quantity of matter called the control mass system. A control mass system is an identified quantity of matter, which may change shape, position, and thermal condition, with time or space or both, but must always entail the same matter.

For the second case, a definite volume called control volume is designated in space, and the boundary of this volume is known as control surface. The amount and identity of the matter in the control volume may change with time, but the shape of the control volume is fixed, that is, the control volume may change its position in time or space or both, but its shape is always preserved.

4.3.2 Integral and Differential Analysis

The analysis in which large control volumes are used to obtain the aggregate forces or transfer rates is termed integral analysis. When the analysis is applied to individual points in the flow field, the resulting equations are differential equations, and the method is termed differential analysis.

4.4 Basic Equation

To simplify the discussions, let us assume the

时，可以采用如下应用模式中的一种。

（1）无论一个还是所有给定的质量微元的运动都必须满足基本定律和相关的补充定律。

（2）无论一个还是所有微元体积在空间中的运动都必须满足基本定律和相关的辅助定律。

在第一种情况中，定律被应用在称为控制质量系统的一定量的物质上。一个控制质量系统是一定量的物质，这些物质能够随着时间或空间或既随时间又随空间而改变形状、位置和热条件，但是系统必须总是包含同样的物质（没有物质溢出或流入系统）。

对于第二种情况，选取的是空间中称为控制体的一定体积，控制体的边界就是控制表面。在控制体中物质的数量和性质可以随时间改变，但是控制体的形状是固定不变的，即控制体可以改变其在时间或空间或者既在时间又在空间中的位置，但总是保持形状不变。

4.3.2 积分分析和微分分析

采用大尺度的控制体来求解集中力或者传输率的分析称为积分分析。当分析是针对流场中单个点进行时，求得的方程是微分方程，这种方法称为微分分析。

4.4 基本方程

为使讨论简化，假设流动是不

flow to be incompressible, that is, the density is treated as invariant. The basic governing equations for an incompressible flow are the continuity and momentum equations.

可压缩的，即认为密度是不变的量。对于不可压缩流动，基本控制方程是连续性方程和动量方程。

4.4.1 Continuity Equation

4.4.1 连续性方程

The continuity equation is based on the conservation of matter. For steady incompressible flow, the continuity equation in differential form is

连续性方程建立在物质守恒基础上。对于定常不可压缩流动，连续性方程的微分形式为

$$\frac{\partial V_x}{\partial x} + \frac{\partial V_y}{\partial y} + \frac{\partial V_z}{\partial z} = 0 \tag{4.9}$$

where V_x, V_y and V_z are the velocity components along x-, y- and z-directions, respectively.

式中，V_x、V_y、V_z 分别是 x、y、z 方向上的速度分量。

Equation (4.9) may also be expressed as

式（4.9）也可以表示为

$$V \cdot \nabla = 0$$

where

式中

$$\nabla \equiv i\frac{\partial}{\partial x} + j\frac{\partial}{\partial y} + k\frac{\partial}{\partial z}$$

and $V = iV_x + jV_y + kV_z$.

且 $V = iV_x + jV_y + kV_z$。

4.4.2 Momentum Equation

4.4.2 动量方程

The momentum equation, which is based on Newton's second law, represents the balance between various forces acting on a fluid element, namely,

动量方程建立在牛顿第二定律基础上，表示作用在一个流体微元上的各种力之间的平衡。这些力是：

(1) Force due to rate of change of momentum, generally referred to as inertia force.

(1) 由动量变化率产生的力，通常指惯性力。

(2) Body forces such as buoyancy force, magnetic force and electrostatic force.

(2) 体积力，诸如浮力、磁力、静电力。

(3) Pressure force.

(3) 压差力。

(4) Viscous forces (causing shear stress).

(4) 黏性力（产生剪切应力）。

For a fluid element under equilibrium, by Newton's second law, we have the momentum equation as

对处于平衡状态的流体微元，由牛顿第二定律，我们推得动量方程为

Inertia force（惯性力）+ Body force（体积力）
+ Pressure force（压力）+ Viscous force（黏性力）= 0 (4.10)

For a gaseous medium, body forces are

对于气态介质，相比于其他

negligibly small compared to other forces and hence can be neglected. For steady incompressible flows, the momentum equation can be written as

力，体积力微小到可被忽略。对于定常不可压缩流动，动量方程可以写为

$$V_x\frac{\partial V_x}{\partial x}+V_y\frac{\partial V_x}{\partial y}+V_z\frac{\partial V_x}{\partial z}=-\frac{1}{\rho}\frac{\partial p}{\partial x}+\nu\left(\frac{\partial^2 V_x}{\partial x^2}+\frac{\partial^2 V_x}{\partial y^2}+\frac{\partial^2 V_x}{\partial z^2}\right) \quad (4.11a)$$

$$V_x\frac{\partial V_y}{\partial x}+V_y\frac{\partial V_y}{\partial y}+V_z\frac{\partial V_y}{\partial z}=-\frac{1}{\rho}\frac{\partial p}{\partial y}+\nu\left(\frac{\partial^2 V_y}{\partial x^2}+\frac{\partial^2 V_y}{\partial y^2}+\frac{\partial^2 V_y}{\partial z^2}\right) \quad (4.11b)$$

$$V_x\frac{\partial V_z}{\partial x}+V_y\frac{\partial V_z}{\partial y}+V_z\frac{\partial V_z}{\partial z}=-\frac{1}{\rho}\frac{\partial p}{\partial z}+\nu\left(\frac{\partial^2 V_z}{\partial x^2}+\frac{\partial^2 V_z}{\partial y^2}+\frac{\partial^2 V_z}{\partial z^2}\right) \quad (4.11c)$$

Equations (4.11a), (4.11b), (4.11c) are the x, y, z components of momentum equation, respectively. These equations are generally known as Navier-Stokes equations. They can also be written as

式（4.11a）、式（4.11b）、式（4.11c）分别是动量方程在 x、y、z 方向上的分量。这些方程就是众所周知的纳维-斯托克斯方程组，也可写为

$$\frac{\mathrm{D}V}{\mathrm{D}t}=-\frac{1}{\rho}\nabla p+\nu\nabla^2 V \quad (4.12)$$

Navier-Stokes equations are nonlinear partial differential equations and there exists no known analytical method to solve them. This poses a major problem in fluid flow analysis. However, the problem is tackled by making some simplifications to the equation, depending on the type of flow to which it is to be applied. For certain flows, the equation can be reduced to an ordinary differential equation of a simple linear type. For some other type of flows, it can be reduced to a nonlinear ordinary differential equation. For the above types of Navier-Stokes equation governing special category of flows, such as potential flow, fully developed flow in a pipe or channel, and boundary layer over flat plates, it is possible to obtain analytical solutions.

纳维-斯托克斯方程是非线性偏微分方程组，还无法用任何已知的解析方法进行求解，因此成为流体流动分析中的一个主要问题。然而，依据将要求解的流动类型，通过对方程做一些简化，这个问题得到了解决。对于某些流动，方程可以简化为一个简单的线性常微分方程。对于其他一些类型的流动，方程也可以被简化为一个非线性的常微分方程。对于满足上述纳维-斯托克斯方程的特殊类型流动，例如势流、在管道或渠道中充分发展的流动、平板的边界层等，我们都可以求得解析解。

By integrating all the terms in the momentum equations above along a stream filament and considering the body force due to the gravity, then we can obtain a simplified Navier-Stokes equation, which is known as Bernoulli equation.

如果对上述动量方程中所有项沿一条流线积分同时考虑由重力引起的体积力，就能够得到一个简化的纳维-斯托克斯方程，即伯努利方程。

$$\frac{p}{\rho} + g \cdot z + \frac{V^2}{2} = \text{constant}（常数） \tag{4.13}$$

Bernoulli equation is the simplification of Navier-Stokes equations for the one-dimensional treatment of inviscid flow. If a flow is inviscid, by considering the body force, the momentum equation for the flow can be expressed as	伯努利方程是一维无黏流体纳维-斯托克斯方程的简化形式。 如果流动是无黏的，同时考虑体积力，流动的动量方程可表示为

$$\frac{DV}{Dt} = F - \frac{1}{\rho}\nabla p \tag{4.14}$$

This equation is known as Eular equation, which is also a simplified momentum equation for an inviscid flow.	这个方程就是欧拉方程，也是简化的无黏流体动量方程。

4.4.3 Equation of State | ### 4.4.3 状态方程

For air at normal temperature and pressure, the density ρ, pressure p and temperature T are connected by the relation $p = \rho R T$, where R is a constant called gas constant. This is known as the thermal equation of state. At high pressures and low temperatures, the above equation of state breaks down. At normal pressures and temperatures, the mean distance between molecules and the potential energy arising from their attraction can be neglected. The gas behaves like a perfect gas or ideal gas in such a situation. At this stage, it is essential to pay attention to the difference between the ideal and perfect gases.

Real gases below critical pressure and above the critical temperature tend to obey the perfect-gas law. The perfect-gas law encompasses both Charles' law and Boyle's law. Charles' law states that, at constant pressure the volume of a given mass of gas varies directly as its absolute temperature. Boyle's law (isothermal law) states that, for constant temperature the density varies directly as the absolute pressure.

对于常温常压下的空气，通过关系式 $p = \rho R T$ 把密度 ρ、压力 p 和温度 T 联系起来，其中 R 是一个常数，称为气体常数，这个关系式就是热状态方程。上述状态方程在高压低温时不成立。在常温常压下，可以忽略分子间的平均距离和由分子间的吸引而产生的势能。在这种情况下，气体表现为完全气体或理想气体。此处，有必要注意理想气体和完全气体的不同之处。

低于临界压力、高于临界温度的真实气体会遵循完全气体定律。完全气体定律包含查理定律和玻意耳定律。查理定律指出，在压力不变时，一定质量的气体，其体积与绝对温度成正比。玻意耳定律（等温定律）指出，在温度不变时，一定质量的气体，其密度与绝对压力成正比。

4.4.4 Boundary Layer Equation

It is essential to understand the physics of the flow process before reducing the Navier-Stokes equations to any useful form, by making appropriate approximations with respect to the flow. For example, let us examine the flow over an aircraft wing, shown in Figure 4.1.

4.4.4 边界层方程

通过对流动进行适当的近似，从而把纳维-斯托克斯方程简化成任何有用形式之前，搞清楚流动过程的物理本质是十分必要的。例如，让我们来分析一个飞机机翼上的流动，如图 4.1 所示。

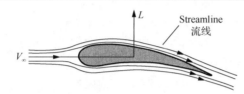

Figure 4.1 Flow past a wing

图 4.1 机翼绕流

This kind of problem is commonly encountered in fluid mechanics. Air flow over the wing creates higher pressure at the bottom, compared to the top surface. Hence, there is a net resultant force component normal to the freestream flow direction, called lift, L, acting on the wing. The velocity varies along the wing chord as well as in the direction normal to its surface. The former variation is due to the shape of the aerofoil, and the later is due to the no-slip condition at the wall. In the direction normal to wing surface, the velocity gradients are very large in a thin layer adjacent to the surface and the flow reaches asymptotically the freestream velocity within a short distance, above the surface. This thin region adjacent to the wall, where the velocity increases from zero to freestream value, is known as the boundary layer. Inside the boundary layer the viscous forces are predominant. Further, it so happens that the static pressure outside the boundary layer, acting in the direction normal to the surface, is transmitted to the boundary through the boundary layer, without appreciable change. In other words, the pressure gradient across the boundary layer is zero. Neglecting

这类问题是流体力学中经常会遇到的问题。流过机翼的气流在（机翼）下部产生了高于上表面的压力。因此，存在一个垂直于来流流动方向、作用在机翼上的合力分量，称为升力 L。速度沿机翼的翼弦方向是变化的，在机翼表面的法线方向也是一样。前者的变化是由机翼的形状引起的，后者是因为壁面的无滑移条件。在机翼表面的法线方向上，邻近表面的一薄层内速度梯度非常大，且在表面上方，流动在短距离内渐近地达到来流速度。这个紧挨壁面速度从零增加到来流速度值的薄层区域就是边界层。在边界层内黏性力占主导地位，而且边界层外部作用在壁面法线方向上的静压通过边界层几乎不变地传递到壁面。换句话说，边界层内的压力梯度为零。忽略流线之间的层间摩擦力，在边界层

the inter-layer friction between the streamlines, in the region outside the boundary layer, it is possible to treat the flow as inviscid. For Inviscid flow, the Navier-Stokes equation can be simplified to become linear. It is possible to obtain the pressures in the field outside the boundary layer and treat this pressure to be invariant across the boundary layer, that is, the pressure in the freestream is impressed through the boundary layer. For low-viscous fluids such as air, we can assume, with a high degree of accuracy, that the flow is frictionless over the entire flow field, except for a thin region near solid surfaces. In the vicinity of solid surface, owing to high velocity gradients, the frictional effects become significant. Such regions near solid boundaries, where the viscous effects are predominant, are termed boundary layers.

In general, boundary layer over streamlined bodies are extremely thin. There may be laminar and turbulent flow within the boundary layer, and its thickness and profile may change along the direction of the flow. Consider the flow over a flat plate shown in Figure 4.2. Different zones of boundary layer over a flat plate are shown in Figure 4.2. The laminar sublayer is that zone adjacent to the boundary, where the turbulence is suppressed to such a degree that only the laminar effects prevail. The various regions shown in Figure 4.2 are not sharp demarcations of different zones. There is actually a gradual transition from one region, where certain effect predominates, to another region, where some other effect is predominant.

外部区域，可以将流动视为无黏的。对于无黏流动，纳维-斯托克斯方程可以简化成线性的方程。边界层外部流场中压力是可以求得的，且这一压力被认为在边界层内沿壁面法向是不变的，即来流中压力决定了边界层中的压力。对于低黏度流体例如空气，除了靠近固体表面的一个薄层区域，我们可以很精确地假设，在整个流场中流动是无摩擦的。但在固体表面四周，由于存在高的速度梯度，摩擦作用显著。邻近固体边界的黏性作用主导的区域称为边界层。

通常，流线体上边界层很薄，边界层里可能存在层流和紊流，且它的厚度及形状会沿着流动方向变化。取一个如图 4.2 所示平板上的流动，图 4.2 中给出了平板上边界层的不同区域。层流底层是紧邻边界的区域，那里紊流被抑制，所以只体现为层流效应。图 4.2 中不同区域之间没有明显的分界，实际上从一种作用占主导到另一种作用占主导的区域之间有个过渡区域。

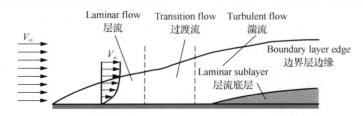

Figure 4.2　Different zones of flat plate boundary layer

图 4.2　平板边界层的不同区域

Although the boundary layer is thin, it plays a vital role in fluid dynamics. The drag on ships, aircraft and missiles, the efficiency of compressors and turbines of jet engines, the effectiveness of ram jets and turbojets, and the efficiencies of numerous other engineering devices are all influenced by the boundary layer to a significant extent. The performance of a device depends on the behavior of boundary layer and its effect on the main flow.

4.5 Rotational and Irrotational Motion

4.5.1 Circulation and Vorticity

When a fluid element is subjected to a shearing force, a velocity gradient is produced perpendicular to the direction of shear, that is, a relative motion occurs between two layers. To encounter this relative motion, the fluid elements have to undergo rotation. A typical example of this type of motion is the motion between two roller chains rubbing each other, but moving at different velocities. It is convenient to use an abstract quantity called circulation, $\varGamma$, defined as the line integral of velocity vector between any two points (to define rotation of the fluid element) in a flow field. By definition,

$$\varGamma = \oint_C V \cdot \mathrm{d}l \tag{4.15}$$

where $\mathrm{d}l$ is an elemental length, C is the path of integration.

Circulation per unit area is known as vorticity ζ,

$$\zeta = \varGamma / A \tag{4.16}$$

In vector form, ζ becomes

$$\zeta = \nabla \times V = \mathrm{curl} V \tag{4.17}$$

where V is the flow velocity, given by $V = iV_x +$

jV_y, and	jV_y，且

$$\nabla \equiv i\frac{\partial}{\partial x} + j\frac{\partial}{\partial y}$$

For a two-dimensional flow in xy-plane, vorticity ζ becomes	对于 xy 平面内的二维流动，涡量 ζ 变成

$$\zeta_z = \frac{\partial V_y}{\partial x} - \frac{\partial V_x}{\partial y} \qquad (4.18a)$$

where ζ_z is the vorticity about the z-direction, which is normal to the flow field. Likewise, the other components of vorticity about x- and y-directions are	式中，ζ_z 是垂直于流场的 z 方向涡量。同样，其他关于 x 方向和 y 方向的涡量分量为

$$\zeta_x = \frac{\partial V_z}{\partial y} - \frac{\partial V_y}{\partial z} \qquad (4.18b)$$

$$\zeta_y = \frac{\partial V_x}{\partial z} - \frac{\partial V_z}{\partial x} \qquad (4.18c)$$

Circulation V can be related to vorticity ζ:	环量 V 可与涡量 ζ 建立关系：

$$\zeta = \nabla \times V$$

by Stokes' theorem:	则由斯托克斯定理：

$$\Gamma = \oint_C V \cdot dl = \iint_S \zeta \cdot dS$$

only if the integration path is a boundary of a closed surface S, not just a closed curve. Thus vorticity is the circulation per unit area, taken around an infinitesimal loop. Correspondingly, the flux of vorticity is the circulation.

If $\zeta = 0$, the flow is known as irrotational flow. Inviscid flows are basically irrotational flows.

4.5.2 Stream Function

Streamlines are imaginary lines in the flow field such that the velocity at all points on these lines are always tangential to them. Flows are usually depicted graphically with the aid of streamlines. Streamlines proceeding through the periphery of an infinitesimal area at some time t forms a tube called streamtube. From the definition of streamlines, it can be inferred that,

(1) Flow cannot cross a streamline, and the mass flow between two streamlines is conserved.

只有当积分路径是一个封闭表面 S 的边界线，而不只是一个封闭曲线时才成立。因此，涡量是无限小封闭环所包围的单位面积上的环量。相应地，涡通量就是环量。

如果 $\zeta=0$，流动就是无旋流动。无黏流动基本上都是无旋流动。

4.5.2 流函数

流线是流体流动中假想的线，流线上各点的流动速度总是与流线相切。通常借助流线来描绘流动。流线在某一瞬时 t 穿过一个无限小面积周边的流线形成一个称为流管的管。由流线的定义，可以推断出：

（1）流动不能穿过流线，且在两流线之间质量流量守恒。

(2) Based on the streamline concept, a function ψ called stream function can be defined. The velocity components of a flow field can be obtained by differentiating the stream function.

In terms of stream function ψ, the velocity components of a two-dimensional incompressible flow are given as

$$V_x = \frac{\partial \psi}{\partial y}, \quad V_y = -\frac{\partial \psi}{\partial x} \tag{4.19}$$

If the flow is compressible the velocity components become

$$V_x = \frac{1}{\rho}\frac{\partial \psi}{\partial y}, \quad V_y = -\frac{1}{\rho}\frac{\partial \psi}{\partial x} \tag{4.20}$$

It is important to note that, the stream function is defined only for two dimensional flows, and the definition does not exist for three-dimensional flows. Even though, some books define ψ for axisymmetric flow, they again prove to be equivalent to two-dimensional flow. We must realize that, the definition of ψ does not exist for three-dimensional flows, because such a definition demands a single tangent at any point on a streamline, which is not possible in three-dimensional flows.

4.5.3 Relationship Between Stream Function and Velocity Potential

For irrotational flows (the fluid elements in the field are free from rotation), there exists a function ϕ called velocity potential or potential function. For a steady two-dimensional flows, ϕ must be a function of two space coordinates (say, x and y). The velocity components are given by

$$V_x = \frac{\partial \phi}{\partial x}, \quad V_y = \frac{\partial \phi}{\partial y} \tag{4.21}$$

From Equations (4.19) and (4.21), we can write

$$\frac{\partial \psi}{\partial y} = \frac{\partial \phi}{\partial x}, \quad \frac{\partial \psi}{\partial x} = \frac{\partial \phi}{\partial y} \tag{4.22}$$

These relations between stream function and potential function, given by Equation (4.22), are the famous Cauchy-Riemann equations of complex variable theory. It can be shown that, the lines of constant ϕ or potential lines form a family of curves which intersect the streamlines in such a manner as to have the tangents of the respective curves always at right angles at the point of intersection. Hence, the two sets of curves given by ψ = constant and ϕ = constant form an orthogonal grid system or flow-net. That is, the streamlines and potential lines in flow field are orthogonal.

Unlike stream function, potential function exists for three-dimensional flows also, because there is no condition like the local flow velocity which must be tangential to the potential lines imposed in the definition of ϕ. The only requirement for the existence of ϕ is that the flow must be potential.

式（4.22）给出的流函数和势函数之间的关系就是复变理论中著名的柯西-黎曼方程。可见，ϕ 等值线或等势线总是在横截处与曲线的切线形成一簇横截流线的曲线，使得曲线的切线在横截点处总是与流线垂直。因此，由 ψ = 常数和 ϕ = 常数给出的两组曲线形成了一个正交的网格系统，或称流网，即流线和等势线在流场中是正交的。

不同于流函数，因为在 ϕ 的定义中，没有像流动速度必须与等势线相切这样的前提，所以势函数也适用于三维流动。ϕ 存在的唯一要求就是流动必须是有势的。

4.6　Potential Flow

Potential flow is based on the concept that the flow field can be represented by a potential function ϕ such that,

4.6　位势流

位势流（势流）的思想在于流场能够被一个势函数 ϕ 表示，因此

$$\nabla^2 \phi = 0 \tag{4.23}$$

This linear partial differential equation is popularly known as Laplace equation. Derivatives of ϕ with respect to the space coordinates x, y and z give the velocity components V_x, V_y and V_z, respectively, along x-, y- and z-directions. Unlike the stream function ψ, the potential function can exist only if the flow is irrotational, that is, when viscous effects are absent. All inviscid flows must satisfy the irrotationality condition,

这个线性偏微分方程就是众所周知的拉普拉斯方程。ϕ 关于空间坐标 x、y、z 的导数分别给出沿 x、y、z 方向的速度分量 V_x、V_y、V_z。不同于流函数 ψ，势函数仅在流动是无旋时，即无黏性作用时才存在。所有的无黏流动一定满足无旋条件，

$$\nabla \times V = 0 \tag{4.24}$$

For two-dimensional potential flows, by Equation (4.18), we have the vorticity ζ as

$$\zeta_z = \frac{\partial V_y}{\partial x} - \frac{\partial V_x}{\partial y} = 0$$

Using Equation (4.21), we get the vorticity as

$$\zeta_z = \frac{\partial^2 \phi}{\partial x \partial y} - \frac{\partial^2 \phi}{\partial x \partial y} = 0$$

This shows that the flow is irrotational. For two-dimensional incompressible flows, the continuity equation is

$$\frac{\partial V_x}{\partial x} + \frac{\partial V_y}{\partial y} = 0$$

In terms of the potential function ϕ, this becomes

$$\frac{\partial^2 \phi}{\partial x^2} - \frac{\partial^2 \phi}{\partial y^2} = 0$$

that is,

$$\nabla^2 \phi = 0$$

This linear equation is the governing equation for potential flows.

For potential flows, the Navier-Stokes equations [Equation (4.11)] reduce to

$$V_x \frac{\partial V_x}{\partial x} + V_y \frac{\partial V_x}{\partial y} + V_z \frac{\partial V_x}{\partial z} = -\frac{1}{\rho} \frac{\partial p}{\partial x} \tag{4.25a}$$

$$V_x \frac{\partial V_y}{\partial x} + V_y \frac{\partial V_y}{\partial y} + V_z \frac{\partial V_y}{\partial z} = -\frac{1}{\rho} \frac{\partial p}{\partial y} \tag{4.25b}$$

$$V_x \frac{\partial V_z}{\partial x} + V_y \frac{\partial V_z}{\partial y} + V_z \frac{\partial V_z}{\partial z} = -\frac{1}{\rho} \frac{\partial p}{\partial z} \tag{4.25c}$$

Equation (4.25) is known as Euler's equation.

At this stage, it is natural to have the following doubts about the streamline and potential function, because we define the streamline as an imaginary line in a flow filed and potential function as a mathematical function, which exists only for inviscid flows. The answers to these vital doubts are the following.

(1) Among the graphical representation concepts

对于二维势流,由式(4.18),我们有涡量 ζ 如下:

利用式(4.21),得到涡量为

这表明流动是无旋的。对于二维不可压缩流动,连续性方程为

以势函数 ϕ 形式,上式可以写成

即

这个线性方程就是势流的控制方程。

对于势流,纳维-斯托克斯方程[式(4.11)]简化为

方程(4.25)就是欧拉方程。

因为我们定义流线为流场中假想的线,势函数是仅适用于无黏流动的数学函数,此处自然会有关于流线和势函数的疑问。这些疑问的回答如下。

(1)在图形化描述的概念

namely, the pathline, streakline and streamline, only the first two are physical, and the concept of streamline is only hypothetical. But even though imaginary, the streamline is the only useful concept. Because, it gives a mathematical representation for the flow field in terms of stream function ψ, with its derivatives giving the velocity components. Once the velocity components are known, the resultant velocity, its orientation, the pressure and temperature associated with the flow can be determined. Thus, streamline plays a dominant role in the analysis of fluid flow.

(2) Knowing pretty well that, no fluid is inviscid or potential, we introduce the concept of potential flow, because this gives rise to the definition of potential function. The derivative of potential function with respective to the spacial coordinates gives the velocity components in the direction of the respective coordinates and the substitution of these velocity components in the continuity equation results in Laplace equation. Even though this equation is the governing equation for an impractical or imaginary flow (inviscid flow), the fundamental solutions of Laplace equation forms the basis for both experimental and computation flow physics. The basic solutions for the Laplace equation are the uniform flow, source, sink, free or potential vortex. These solutions being potential, can be superposed to get the mathematical functions representing any practical geometry of interest. For example, superposition of a doublet and uniform flow would represent the flow past a circular cylinder. In the same manner, suitable distribution of source and sink along the camber line and superposition of uniform flow over this distribution will mathematically represent the flow past an aerofoil. Thus, any practical geometry can be modeled mathematically, using the basic solutions of the Laplace equation.

中，即迹线、脉线和流线中，只有前两个是实际的线，流线的概念仅是假想的。尽管是假想的，流线的概念却是唯一有用的概念。因为它以流函数 ψ 的形式给出了流场的数学描述。流函数是对流线微分给出的速度分量，一旦速度分量已知，速度及其方向和与流动相关的压力和温度就可以确定了。所以，流线在流体流动分析中起着至关重要的作用。

（2）尽管对这一点知道得很清楚，即流体都不是无黏的或者是有势的，但我们仍然引入势流这个概念，因为这样引出了势函数的定义。势函数对空间坐标的导数给出了各坐标方向上的速度分量，将这些速度分量代入连续性方程可推导出拉普拉斯方程。尽管这个方程是一个非实际的或假想（无黏的）流动的控制方程，拉普拉斯方程的基本解还是构成了实验和计算流体力学的基础。拉普拉斯方程的基本解是均匀流、源、汇、自由涡流或者有势涡流。这些有势的解，可以叠加起来得到描述许多想要求解的实际形状流动的数学方程。例如，一个偶极子和均匀流的叠加将表示一个圆柱绕流的流动。采用同样的方式，源和汇沿脊线的适当分布，以及在这个分布上叠加均匀流，可定量地描述翼型绕流流动。因此采用拉普拉斯方程的基本解，可以建立具有任何实际几何形状物体绕流的数学模型。

4.6.1 Two-Dimensional Source and Sink

Source is a potential flow field in which flow emanating from a point spreads radially outwards, as shown in Figure 4.3(a). Sink is potential flow field in which flow gushes towards a point from all radial directions, as illustrated in Figure 4.3(b).

4.6.1 二维的源和汇

源是这样一种有势流场，在这一流场中，流动从一个点径向地向外传播，如图 4.3（a）所示。汇是这样一种有势流场，在这一流场中，流动从所有径向方向涌向一点，如图 4.3（b）所示。

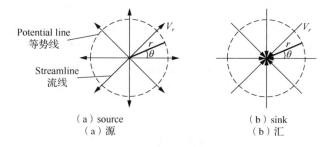

Figure 4.3 Illustration of two-dimensional source and sink

图 4.3 二维源和汇的示意图

Consider a source at origin, shown in Figure 4.3(a). The volume flow rate $\dot{q}$ crossing a circular surface of radius r and unit depth is given by

考察位于原点的一个源，如图 4.3（a）所示。流过一个半径为 r 的单位厚度圆柱表面的体积流量 $\dot{q}$ 为

$$\dot{q} = 2\pi r V_r \tag{4.26}$$

where V_r is the radial component of velocity. The volume flow rate $\dot{q}$ is referred to as the strength of the source. For a source, the radial lines are streamlines. Therefore, the potential lines must be concentric circles, represented by

式中，V_r 是径向速度分量。体积流量 $\dot{q}$ 是源的强度。对于一个源，径向线是流线，因此，等势线就是同心圆，表示为

$$\phi = A \ln(r) \tag{4.27}$$

where A is a constant. The radial velocity component $V_r = \partial \phi / \partial r = A/r$.

Substituting this into Equation (4.27), we get

式中，A 是常数。径向速度分量 $V_r = \partial \phi / \partial r = A/r$。

将其代入式（4.26）可得

$$\frac{2\pi r A}{r} = \dot{q}$$

or

或者

$$A = \frac{\dot{q}}{2\pi}$$

Thus, the velocity potential for a two-dimensional source of strength $\dot{q}$ becomes	因此，一个强度为$\dot{q}$的二维源速度势变成

$$\phi = \frac{\dot{q}}{2\pi}\ln(r) \tag{4.28}$$

In a similar manner as above, the stream function for a source of strength $\dot{q}$ can be obtained as	同上，一个强度为$\dot{q}$的源，其流函数为

$$\psi = \frac{\dot{q}}{2\pi}\theta \tag{4.29}$$

where θ is the orientation (inclination) of the streamline from the *x*-direction, measured in the counter clockwise direction, as shown in Figure 4.3(a). Similarly, for a sink, which is a type of the flow in which the fluid from infinity flows radially towards the origin, we can show that, the potential and stream functions are given by	式中，θ是逆时针方向测量时流线距离（偏离）*x*方向的角度，如图4.3（a）所示。同样，对于一个汇，即一种流体从无限远处径向地流入原点的流动，我们可以给出势函数和流函数为

$$\phi = -\frac{\dot{q}}{2\pi}\ln(r)$$

and	和

$$\psi = -\frac{\dot{q}}{2\pi}\theta$$

where $\dot{q}$ is the strength of the sink. Note that the volume flow rate is termed the strength of the sink. Also, for both source and sink the origin is a singular point.	式中，$\dot{q}$是汇的强度。注意，体积流量被称为源和汇的强度。还有，对于源和汇，原点都是一个奇异点。

4.6.2 Simple Vortex

4.6.2 简单涡

A simple or free vortex is a flow field in which the fluid elements simply move along concentric circles, without spinning about their own axes. That is, the fluid elements have only translatory motion in a free vortex. In addition to moving along concentric paths, if the fluid elements spin about their own axes, the flow is termed forced vortex.	一个简单或自由涡是流体微元在其中只沿着同心圆运动的流场，流体微元没有绕自身轴的旋转。即在一个简单涡中流体微元仅有平移运动。除了沿着同心路径运动外，如果流体微元还绕自身轴旋转，这种流动称作受迫涡。
A simple or free vortex can be established by selecting the stream function, ψ, of the source to be the potential function ϕ of the vortex. Thus, for a simple vortex	通过选择源的流函数ψ，构成涡的势函数ϕ，可以构建一个简单涡或者自由涡。那么，对于简单涡，有

$$\phi = \frac{\dot{q}}{2\pi}\theta \tag{4.30}$$

| It can be easily shown from Equation (4.30) that, the stream function for a simple vortex is | 从式（4.30）很容易看出，一个简单涡的流函数为 |

$$\psi = \frac{\dot{q}}{2\pi}\ln(r) \qquad (4.31)$$

| It follows from Equations (4.30) and (4.31) that, the velocity components of the simple vortex, shown in Figure 4.4, are | 如图 4.4 所示，由式（4.30）和式（4.31）可得简单涡的速度分量为 |

$$V_\theta = \frac{\dot{q}}{2\pi r}, \quad V_r = 0 \qquad (4.32)$$

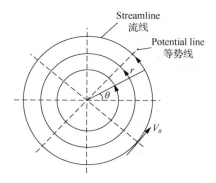

Figure 4.4　A simple or potential vortex flow

图 4.4　一个简单的或有势的涡

| Here again the origin is a singular point, where the tangential velocity V_θ tends to be the infinity, as seen from Equation (4.32). The flow in a simple or free vortex resembles part of the common whirlpool found while we paddle a boat or while empty water from a bathtub. An approximate profile of a whirlpool is as shown in Figure 4.5. | 由式（4.32）可以看出，这里原点也是一个切向速度 V_θ 趋于无穷的奇异点。简单涡或者自由涡流动类似划船或排空浴缸水时常见到的旋涡。一个旋涡的近似轮廓如图 4.5 所示。 |

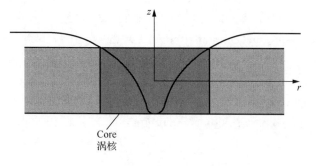

Figure 4.5　A whirlpool flow field

图 4.5　旋涡流场

For the whirlpool, shown in Figure 4.5, the circulation along any path about the origin is given by,	对于图 4.5 中所示的这个旋涡，在原点附近沿任意路径的环量可写为

$$\Gamma = \oint V \cdot \mathrm{d}l$$
$$= \int_0^{2\pi} V_\theta r \mathrm{d}\theta$$

By Equation (4.32), $V_\theta = \dot{q}/2\pi r$, therefore, the circulation becomes	由式（4.32），有 $V_\theta = \dot{q}/2\pi r$，因此，环量变为

$$\Gamma = \int_0^{2\pi} \frac{\dot{q}}{2\pi r} r \mathrm{d}\theta = \dot{q}$$

Since there are no other singularities for the whirlpool, shown in Figure 4.5, this must be the circulation for all paths about the origin. Consequently, $\dot{q}$ in the case of vortex is the measure of circulation about the origin and is also referred to as the strength of the vortex.

因为这个涡不存在其他的奇异点，如图 4.5 所示，这一定是绕原点的所有路径的环量。因而，就涡流来说 $\dot{q}$ 是关于原点的环量的度量，也被称为涡流的强度。

4.6.3 Source-Sink Pair

This is a combination of a source and sink of equal strength, situated (located) at a distance apart. The stream function due to this combination is obtained simply by adding the stream functions of the source and sink. When the distance between the source and sink is made negligibly small, in the limiting case, the combination results in a doublet.

4.6.4 Doublet

A doublet or a dipole is a potential flow field due to a source and sink of equal strength, brought together in such a way that the product of their strength and the distance between them remain constant. Consider a point P in the field of a doublet formed by a source and a sink of strength $\dot{q}$ and $-\dot{q}$, kept at a distance $\mathrm{d}s$, as shown in Figure 4.6, with sink at the origin.

4.6.3 源-汇对

这是一个间隔一定距离放置的等强度源和汇的组合。仅通过源和汇流函数的叠加就可得到这个组合的流函数。当源和汇之间的距离小到微乎其微时，在这种极限情况下，这个组合就形成偶极子。

4.6.4 偶极子

偶极子或双极子是把等强度的源和汇，通过保持其强度与间距乘积不变的方式放置在一起而形成的有势流场。由强度为 $\dot{q}$ 的源和 $-\dot{q}$ 的汇组成的一个偶极子，保持源和汇距离 $\mathrm{d}s$ 且汇在原点，取流场中的一点 P，如图 4.6 所示。

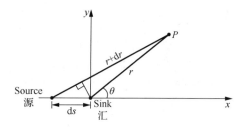

Figure 4.6 Source and sink

图 4.6 源和汇

By Rankine's theorem, the velocity potential of the doublet, ϕ_D, can be expressed as the sum of the velocity potentials of the source and sink. Thus, we have

由兰金理论可知，偶极子的速度势 ϕ_D 可表示为源和汇速度势之和。于是有

$$\phi_D = \frac{\dot{q}}{2\pi}\ln(r+dr) - \frac{\dot{q}}{2\pi}\ln(r)$$
$$= \frac{\dot{q}}{2\pi}\ln\left(\frac{r+dr}{r}\right)$$
$$= \frac{\dot{q}}{2\pi}\ln\left(1+\frac{dr}{r}\right)$$

Expanding $\ln\left(1+\dfrac{dr}{r}\right)$, we get

展开 $\ln\left(1+\dfrac{dr}{r}\right)$，得到

$$\ln\left(1+\frac{dr}{r}\right) = \frac{dr}{r} - \frac{1}{2}\left(\frac{dr}{r}\right)^2 + \cdots$$

But $\dfrac{dr}{r} \ll 1$, therefore, neglecting the second and higher order terms, we get the potential function for a doublet as

但是 $\dfrac{dr}{r} \ll 1$，因此，忽略第二项和高阶项，我们得到一个偶极子的势函数为

$$\phi_D = \frac{\dot{q}}{2\pi}\frac{dr}{r}$$

By the definition of doublet, $ds \to 0$, therefore,

由偶极子的定义得出，$ds \to 0$，因此

$$dr = ds\cos\theta$$

Hence,

于是，

$$\phi_D = \frac{\dot{q}}{2\pi r}ds\cos\theta$$

Also, for a doublet, by definition, $\dot{q}ds =$ constant. Let this constant, known as the strength of the doublet be denoted by m, then

同样，对于一个偶极子，由定义可知，$\dot{q}ds =$ 常数。把这个偶极子强度的常数用 m 来表示，那么

$$m = \dot{q}\mathrm{d}s$$

$$\phi_D = \frac{m}{2\pi r}\cos\theta \tag{4.33}$$

In Cartesian coordinates, the velocity potential for the doublet becomes	在笛卡儿坐标系中，偶极子的速度势为

$$\phi_D = \frac{m}{2\pi}\left(\frac{x}{x^2+y^2}\right)$$

From the above equations for ϕ_D, the expression for the stream function ψ_D can be obtained as	由上述 ϕ_D 的方程能够得到流函数 ψ_D 的表达式为

$$\psi_D = -\frac{m}{2\pi r}\sin\theta$$

In Cartesian coordinates, the stream function becomes	在笛卡儿坐标系中，流函数变成

$$\psi_D = -\frac{m}{2\pi}\left(\frac{y}{x^2+y^2}\right)$$

If the source and sink were placed on the x-axis, the streamlines of the doublet will be as shown in Figure 4.7.	如果源和汇位于 x 轴上，这个偶极子的流线将如图 4.7 所示。

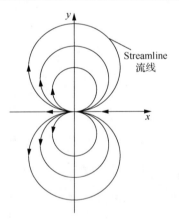

Figure 4.7　Doublet with source and sink on the x-axis

(source located on the left and sink on the right of the origin)

图 4.7　x 轴上由源和汇组成的偶极子（源在原点左侧，汇在原点右侧）

If the source and sink are placed on the y-axis, the resulting expressions for the ϕ_D and ψ_D will become	如果源和汇位于 y 轴上，相应 ϕ_D 和 ψ_D 的表达式将变成

$$\phi_{D(yy)} = \frac{m}{2\pi r}\sin\theta = \frac{m}{2\pi}\left(\frac{y}{x^2+y^2}\right)$$

$$\psi_{D(yy)} = -\frac{m}{2\pi r}\cos\theta$$
$$= -\frac{m}{2\pi}\left(\frac{x}{x^2+y^2}\right)$$

The streamlines of the doublet will be as shown in Figure 4.8. The expression for the stream function can be arranged in the form:	偶极子的流线如图 4.8 所示。流函数的表达式可整理为如下形式：

$$\psi_{D(yy)} = \frac{cx}{x^2+y^2}$$

where $c = m/(2\pi)$, is a constant. This can be expressed as	式中，$c = m/(2\pi)$，是一个常数。方程可表示为

$$x^2 + y^2 + \frac{cx}{\psi_{D(yy)}} = 0$$

$$\left(x + \frac{c}{2\psi_{D(yy)}}\right)^2 + y^2 = \left(\frac{c}{2\psi_{D(yy)}}\right)^2$$

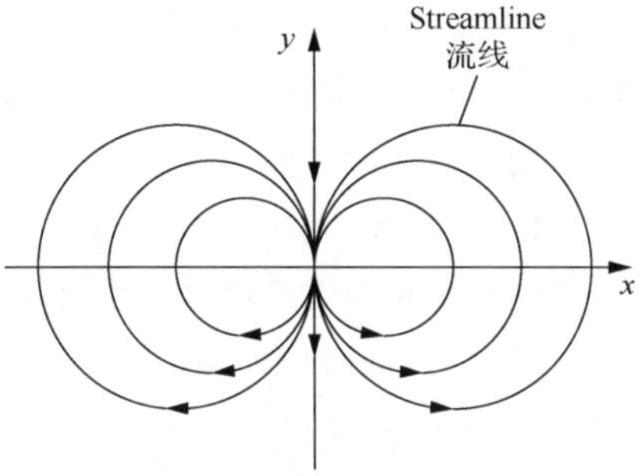

Figure 4.8 Doublet with source and sink on the *y*-axis

(source located below the origin and sink above the origin)

图 4.8 *y* 轴上由源和汇组成的偶极子（源在原点下方，汇在原点上方）

Thus, the stream lines represented by $\psi_{D(yy)}$ = constant are circles with their centers lying on the *x*-axis and are tangent to the *y*-axis at the origin (Figure 4.8). Direction of flow at the origin is along the negative *y*-axis, pointing outward from the source of the limiting source-sink pair, which is called the axis of the doublet.	那么，这些由"$\psi_{D(yy)}$ = 常数"表示的流线，是圆心在 *x* 轴上且与 *y* 轴相切于原点的圆（图 4.8）。在原点处流动方向是沿着 *y* 轴的负方向，从无限趋近的指定源-汇对的源指向外的，称为偶极子的轴。
The potential and stream functions for the concentrated source, sink, vortex, and doublet are all singular at the origin. It will be shown in the following	位于原点的源、汇、涡流和偶极子的势函数及流函数在原点处都是奇异的。后续章节将会

section that several interesting flow patterns can be obtained by superposing an uniform flow on these concentrated singularities.

4.7 Flow Past a Half-Body—Combination of Simple Flows

In Section 4.6 we saw that, flow past practical shapes of interest can be represented or simulated with suitable combination of source, sink, free vortex and uniform flow. In this section let us discuss some such flow fields.

An interesting pattern of flow past a half-body, shown in Figure 4.9, can be obtained by combining a source and an uniform flow parallel to x-axis. By definition, a given streamline (ψ = constant) is associated with one particular value of the stream function. Therefore, if we join the points of intersection of the radial streamlines of the source with the rectilinear streamlines of the uniform flow, where the sum of magnitudes of the two stream functions is equal to the streamline of the resulting combined flow pattern. If this procedure is repeated for a number of values of the combined stream function, the result will be a picture of the combined flow pattern.

The stream function for the flow due to the combination of a source of strength $\dot{q}$ at the origin, immersed in an uniform flow of velocity V_∞, parallel to x-axis, is

$$\psi = V_\infty r \sin\theta + \frac{\dot{q}}{2\pi}\theta \tag{4.34}$$

The streamlines of the resulting flow field will be as shown in Figure 4.9.

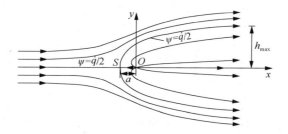

Figure 4.9　Uniform irrotational flow past a two-dimensional half-body

图 4.9　绕过一个二维半体的均匀无旋流动

The streamline passing through the stagnation point S is termed the stagnation streamline. The stagnation streamline resembles a semi-ellipse. This shape is popularly known as Rankine's half-body. The streamlines inside the semi-ellipse are due to the source and those outside the semiellipse are due to the uniform flow. The boundary or stagnation streamline is given by

通过驻点 S 的流线称为滞止流线。滞止流线类似于一个半椭圆。这个形状也就是众所周知的兰金半体。半椭圆内部流线是由源产生的，外部流线源于均匀流。边界或滞止流线由下式给出：

$$\psi = \frac{\dot{q}}{2}$$

It is seen that, S is the stagnation point where the uniform flow velocity V_∞ cancels the velocity of the flow from the source. The stagnation point is located at (a, π). At the stagnation point, both V_r and V_θ should be zero. Thus,

我们可以看到 S 是驻点（滞止点），在该点均匀流的速度 V_∞ 抵消了源的流速。驻点位于 (a, π)，在驻点处，速度 V_r 和 V_θ 都应该为 0。于是，

$$V_r = \frac{1}{r}\frac{\partial \psi}{\partial \theta}$$
$$= V_\infty \cos\theta + \frac{\dot{q}}{2\pi r}$$
$$= 0$$

This gives,　　　　　　　　　　　　　　　所以有

$$V_\infty - \frac{\dot{q}}{2\pi a} = 0$$
$$a = \frac{\dot{q}}{2\pi V_\infty}$$

Therefore, the stream function of the stagnation point is

因此，驻点的流函数为

$$\psi_s = V_\infty r \sin\theta + \frac{\dot{q}\theta}{2\pi}$$

At the stagnation point S, $r=a$ and $\theta=\pi$,

在驻点 S 处，$r=a$，且

therefore,	$\theta = \pi$,因此,
$$\psi_s = V_\infty a \sin\pi + \frac{\dot{q}\pi}{2\pi}$$ $$= \frac{\dot{q}}{2}$$	
The equation of the streamline passing through the stagnation point is obtained by setting $\psi = \psi_s = \dot{q}/2$, resulting in	设 $\psi = \psi_s = \dot{q}/2$,得到驻点的流线方程为
$$V_\infty r \sin\theta + \frac{\dot{q}\theta}{2\pi} = \frac{\dot{q}}{2}$$	
A plot of the streamlines represented by Equation (4.34) is shown in Figure 4.9. It is a semi-infinite body with a smooth nose, generally called a half-body. The stagnation streamline divides the field into a region external to the body and a region internal to it. The internal flow consists entirely of the fluid emanating from the source, and the external region contains the originally uniform flow. The half-body resembles several shapes of theoretical interest, such as the front part of a bridge pier or an aerofoil. The upper half of the flow resembles the flow over a cliff or a side contour of a wide channel. The half-width of the body is given as	图 4.9 所示为式(4.34)描述的流线图。它是一个具有光滑前缘的半无限体,一般称作半体。滞止流线把流场分成了半体的外部区域和内部区域。内部流全部由源发出的流体构成,外部区域包含原来的均匀流。半体与多个理论研究关注的形状相似,例如,桥墩或机翼的前缘。上半部分流动类似于流过峭壁或宽渠侧壁的流动。 半体的半宽可写为
$$h = r\sin\theta = \frac{\dot{q}(\pi - \theta)}{2\pi V_\infty}$$	
As $\theta \to 0$, the half-width tends to be a maximum of $h_{max} = \dot{q}/(2V_\infty)$, that is, the mass flux from the source is contained entirely within the half-body, and $\dot{q} = (2h_{max})V_\infty$ at a large downstream distance where the local flow velocity $u = V_\infty$. The pressure distribution can be found from the incompressible Bernoulli's equation	当 $\theta \to 0$ 时,半宽度趋向于最大值 $h_{max} = \dot{q}/(2V_\infty)$,即来自源的质量流量被全部包含于半体之内。在下游当地流速为 $u = V_\infty$ 的远处, $\dot{q} = (2h_{max})V_\infty$。 由不可压缩伯努利方程可得压力分布为
$$p + \frac{1}{2}\rho u^2 = p_\infty + \frac{1}{2}\rho V_\infty^2$$	
where p and u are the local static pressure and velocity of the flow, respectively. The pressure can be expressed through the non-dimensional pressure difference called the pressure	式中, p 和 u 分别是当地的静压和流速。 我们可利用称为压力系数的无量纲压差来表示压力,

coefficient, defined as | 定义为

$$C_p = \frac{p - p_\infty}{\frac{1}{2}\rho_\infty V_\infty^2}$$

where p and p_∞ are the local and freestream static pressures, respectively, ρ_∞ is the freestream density and V_∞ is freestream velocity.	式中，p 和 p_∞ 分别是当地和来流的静压；ρ_∞ 是来流的密度；V_∞ 是来流的速度。
A plot of the C_p distribution on the surface of the half-body is shown in Figure 4.10. It is seen that, there is a positive pressure or compression zone near the nose of the body and the pressure becomes negative or suction, downstream of the positive pressure zone. This positive pressure zone is also called pressure-hill. The net pressure force acting on the body can easily be shown to be zero, by integrating the pressure p acting on the surface. The half-body is obtained by the linear combination of the individual stream functions of a source and a uniform flow, as per the Rankine's theorem which states that, "the resulting stream function of n potential flows can be obtained by combining the stream functions of the individual flows". This half-body is also referred to as Rankine's half-body.	C_p 在半体表面的分布图如图 4.10 所示，可见靠近物体前缘处存在一个正压区或压缩区。在正压区下游，压力变为负压或吸力。正压区也称为压力丘。对作用在表面的压力 p 积分，很容易得出作用在体上的总压为 0。半体的流函数是通过把一个点源和一个均匀流的流函数进行线性组合获得的，如兰金理论所表述的，"通过组合单个流动的流函数可获得 n 个势流的合域流函数"。这个半体也称为兰金半体。

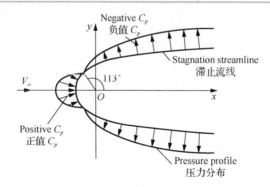

Figure 4.10 Pressure distribution for potential flow over a half-body

图 4.10 绕半体的势流压力分布

Example 4.1	**例 4.1**
A two-dimensional source of strength 4.0m²/s is placed in an uniform flow of velocity 1m/s parallel	一个强度为 4.0m²/s 的二维源置于一个平行于 x 轴、速度为

to *x*-axis. Determine the flow velocity and its direction at $r = 0.8$m and $\theta = 140°$.	1m/s 的均匀流中。求 $r = 0.8$m 且距离 x 轴逆时针 $\theta = 140°$ 处的流速及其方向。

Solution
    The given flow is an ideal flow around a half-body. The stream function for the flow around a half-body, by Equation (4.34), is
	解
	    给定的流动是一个绕半体的理想流动。根据式（4.34），绕半体流动的流函数为

$$\psi = V_\infty r \sin\theta + \frac{\dot{q}}{2\pi}\theta$$

Given that, $V_\infty = 1$m/s and $\dot{q} = 4$m/s. Thus,	已知，$V_\infty = 1$m/s、$\dot{q} = 4$m/s，那么

$$\psi = r\sin\theta + \frac{4}{2\pi}\theta$$

The tangential and radial components of the velocity, respectively, are	速度的切向和径向分量分别为

$$V_\infty = -\frac{\partial \psi}{\partial r}$$
$$= -\sin\theta = -\sin 140°$$
$$= -0.643 \text{m/s}$$

and	和

$$V_r = \frac{1}{r}\frac{\partial \psi}{\partial \theta}$$
$$= \frac{1}{r}\left(r\cos\theta + \frac{4}{2\pi}\right)$$
$$= \frac{1}{0.8} \times \left(0.8\cos 140° + \frac{2}{\pi}\right)$$
$$= 0.0297 \text{m/s}$$

The resultant velocity, at $r = 0.8$m and $\theta = 140°$, is	在 $r = 0.8$m、$\theta = 140°$ 处合成的速度为

$$V = \sqrt{V_r^2 + V_\theta^2}$$
$$= \sqrt{0.0297^2 + (-0.643)^2}$$
$$= 0.644 \text{m/s}$$

The velocity field is as shown in the Figure 4.11.	速度场如图 4.11 所示。

Chapter 4 Kinematics and Dynamics of Fluid Flow 流体运动学和动力学

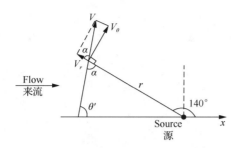

Figure 4.11 Velocity field

图 4.11 速度场

If θ' is the angle the velocity makes with the horizontal, as shown in the figure, then	如果 θ' 是速度与水平线的夹角，如图中所示，那么

$$\theta' = \theta - \alpha$$

Also,	同样

$$\tan\alpha = \frac{V_\theta}{V_r} = \frac{0.643}{0.0297}$$
$$= 21.65$$
$$\alpha = \arctan(21.65)$$
$$= 87.36°$$

Thus,	则

$$\theta = 140° - 87.36°$$
$$= 52.64°$$

Example 4.2 A two-dimensional flow field is made up of a source at the origin and a flow given by $\phi = r^2\cos 2\theta$. Locate any stagnation points in the upper half of the coordinate plane $(0 \leqslant \theta \leqslant \pi)$. **Solution** The potential function of the flow is	**例 4.2** 一个位于原点的源和一个由 $\phi = r^2\cos 2\theta$ 给定的流动构成一个二维流场。确定在上半坐标平面内的所有驻点的位置 $(0 \leqslant \theta \leqslant \pi)$。 **解** 流动的势函数为

$$\phi = \frac{m}{2\pi}\ln r + r^2\cos 2\theta$$

where $\frac{m}{2\pi}\ln r$ is the velocity potential for the source, with strength $m(\mathrm{m}^3/\mathrm{s})$. The velocity components are given by	式中，$\frac{m}{2\pi}\ln r$ 是源的速度势，强度为 $m(\mathrm{m}^3/\mathrm{s})$。 速度分量为

$$V_r = \frac{\partial \phi}{\partial r}$$
$$V_\theta = \frac{1}{r}\frac{\partial \phi}{\partial \theta}$$

$$V_r = \frac{m}{2\pi r} + 2r\cos 2\theta$$

$$V_\theta = -2r\sin 2\theta$$

At the stagnation point, $V_r = 0$ and $V_\theta = 0$. Thus, we have	在滞止点，$V_r = 0$，$V_\theta = 0$。于是有

$$\cos 2\theta = -\frac{m}{4\pi r^2} \qquad (4.35)$$

and	和

$$\sin 2\theta = 0$$

$$2\theta = 0 \text{ or（或）} \pi$$

$$\theta = 0 \text{ or（或）} \frac{\pi}{2}$$

Thus, $\theta_s = \pi/2$ at the stagnation point. Substitution of this into Equation (4.35) gives,	所以在滞止点 $\theta_s = \pi/2$，把它代入式（4.35）中，给出

$$r_s = \left(\frac{m}{4\pi}\right)^{1/2}$$

Example 4.3　　A certain body has the shape of Rankine half-body of the maximum thickness of 0.5m. If this body is to be placed in an air stream of velocity 30m/s, find the source strength required to simulate flow around the body? **Solution**　　The half-body can be represented as a combination of a source and a uniform flow.　　The resulting stream function is	**例 4.3**　　某一兰金半体形状的物体，最大厚度为0.5m。如果这个物体被放置在速度为30m/s的气流里，如何确定能够产生绕流所需源的强度？ **解**　　半体可表示为一个源和一个均匀流的组合。　　合成的流函数为

$$\psi = \psi_{均匀流} + \psi_{源}$$

$$\psi = Ur\sin\theta + \frac{m}{2\pi}\theta$$

where m is the source strength. At the stagnation point S,	式中，m 是源的强度。在驻点 S 处，

$$V_r = \frac{1}{r}\frac{\partial \psi}{\partial \theta} = 0$$

That is,	即

$$U\cos\theta + \frac{m}{2\pi r} = 0$$

But at S, $\theta = \pi$, thus,	但在 S 处，$\theta = \pi$，那么

$$-U + \frac{m}{2\pi r} = 0$$

$$U = \frac{m}{2\pi r}$$

If $x = b$ is the stagnation point, then at $r = b$,	如果 $x = b$ 是驻点,那么在 $r = b$ 处,

$$U = \frac{m}{2\pi r}$$

or	或

$$U = \frac{m}{2\pi b}$$

The value of the stream function of the streamline passing through the stagnation point can be obtained by evaluating Equation (4.34), at $r = b$ and $\theta = \pi$, which yields	通过求解方程(4.34)可获得通过驻点流线的流函数值;当 $r = b$,且 $\theta = \pi$ 时,得

$$\psi_{stag} = \frac{m}{2} = \frac{2\pi b U}{2}$$
$$= \pi b U$$

Thus, from Equation (4.34), we get	那么,根据方程(4.34)得到

$$\pi b U = U r \sin\theta + \frac{m}{2\pi}\theta$$
$$= U r \sin\theta + \frac{\theta}{\pi}\pi b U$$

By solving, we get	求解得到

$$r = \frac{b(\pi - \theta)}{\sin\theta} \tag{4.36}$$

The width of the half-body asymptotically approaches $2\pi b$, as shown below. Equation (4.36) can be written as	半体的宽度逐渐接近 $2\pi b$,如下所示,式(4.36)可写成

$$r\sin\theta = b(\pi - \theta)$$

But $y = r\sin\theta$, thus,	但是, $y = r\sin\theta$,因此

$$y = b(\pi - \theta) \tag{4.37}$$

From Equation (4.37), it is seen that, as $\theta \to 0$ or $\theta \to 2\pi$ the half-width approaches $\pm b\pi$. For $\theta = 0$, Equation (4.37) gives	由式(4.37),可见当 $\theta \to 0$ 或 $\theta \to 2\pi$ 时,半宽度接近 $\pm b\pi$。对于 $\theta = 0$,式(4.37)给出

$$y = b\pi$$

But	但是,

$$b = \frac{m}{2\pi U}$$

Therefore,	因此,

$$y = \frac{m}{2\pi U}\pi$$

or	或

$$m = 2Uy$$

For $U = 30\text{m/s}$ and $y = 0.25\text{m}$, we have,	对于 $U = 30\text{m/s}$、$y = 0.25\text{m}$，我们有

$$m = 2 \times 30 \times 0.25 = 15\text{m}^3/\text{s}$$

The source strength required is $15\text{m}^3/\text{s}$.	所需源的强度为 $15\text{m}^3/\text{s}$。
Example 4.4	**例 4.4**
Check whether the flow represented by the stream function	判断由如下流函数表示的流动是否是无旋的：

$$\psi = V_\infty r \sin\theta + \frac{\dot{q}}{2\pi}\theta$$

where $\dot{q}$ is the volume flow rate, which is a constant, is irrotational.	式中，$\dot{q}$ 是体积流量，为常数。
Solution	**解**
The radial and tangential components of velocity of the given flow are	给定流动的径向和切向速度分量为

$$V_r = \frac{1}{r}\frac{\partial \psi}{\partial \theta} = V_\infty \cos\theta + \frac{\dot{q}}{2\pi r}$$

$$V_\theta = -\frac{\partial \psi}{\partial r} = -V_\infty \sin\theta$$

The irrotationality condition given by Equation (4.24) is	式（4.24）给定的无旋条件为

$$\zeta = \frac{\partial V_y}{\partial x} - \frac{\partial V_x}{\partial y} = 0$$

In terms of r and θ, this becomes	以 r 和 θ 的形式，上式转化为

$$\frac{1}{r}\frac{\partial}{\partial r}(rV_\theta) - \frac{1}{r}\frac{\partial V_r}{\partial \theta} = 0$$

Thus,	于是

$$\frac{1}{r}\frac{\partial}{\partial r}(-rV_\infty \sin\theta) - \frac{1}{r}\frac{\partial}{\partial \theta}\left(V_\infty \cos\theta + \frac{\dot{q}}{2\pi r}\right) = 0$$

$$-\frac{V_\infty \sin\theta}{r} + \frac{1}{r}V_\infty \sin\theta = 0$$

The irrotational condition is satisfied and hence the flow is irrotational.

4.8 Summary

This chapter mainly introduces some basic concepts, analysis methods and basic laws that may relate to the kinematics and dynamics of the fluid flow. In addition, the rotational and irrotational motion of fluid flow are also described and the potential flow was illustrated by giving several simple fluid flow examples.

Basically two treatments are followed for the fluid flow analysis. They are the Lagrangian and Eulerian descriptions. If properties and flow characteristics at each position in the space remain invariant with time, the flow is called the steady flow. A time dependent flow is referred to as unsteady flow. A fluid flow can be described graphically with the pathline, the streakline, streamlines and the timeline. Four basic laws must be satisfied for any continuous medium. They are the conservation of matter (continuity equation), Newton's second law (momentum equation), the conservation of energy (first law of thermodynamics) and increase of the entropy principle (second law of thermodynamics).

4.9 Exercises

Problem 4.1 Air flows through a tube, as shown in Figure 4.12, with a velocity of 0.2m/s and the temperature gradient in the flow direction is $2℃/\text{m}$. The temperature of each particle is increasing at a rate of $0.5℃/\text{s}$ due to the absorption of the thermal radiation. Find the rate of change of the air temperature as recorded by a stationary temperature probe. Also, find whether the temperature field is steady.

[Ans: $0.1℃/\text{s}$, unsteady]

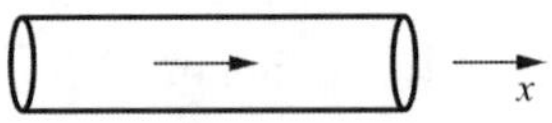

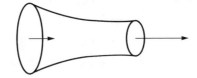

Figure 4.12 Air flow through a tube

图 4.12 流经圆管的气流

Problem 4.2 The flow through the convergent nozzle shown in Figure 4.13 is approximated as one-dimension. If the flow is steady will there be any fluid acceleration? If there is the acceleration, obtain an expression for it in terms of volumetric flow rate $\dot{Q}$ if the area of cross-section is given by $A(x) = e^{-x}$.

[Ans: $\left(\dfrac{\dot{Q}}{e^{-x}}\right)^2$]

题 4.2 流经如图 4.13 所示收缩喷嘴的流动可以近似为一维流动，如果流动是定常的，是否存在流体加速度？如果存在，给出用体积流量 $\dot{Q}$ 表示的加速度表达式，假设过流截面 A 用 $A(x) = e^{-x}$ 表示。

【答：$\left(\dfrac{\dot{Q}}{e^{-x}}\right)^2$】

Figure 4.13 Flow through a convergent nozzle

图 4.13 收缩喷嘴流动

Problem 4.3 Atmospheric air is cooled by a desert cooler by 18℃ and sent into a room. The cooled air then flows through the room and picks up heat from the room at a rate of $0.15℃/s$. The air speed in the room is $0.72\,\text{m/s}$. After some time from switching on, the temperature gradient assumes a value of $0.9℃/m$ in the room. Determine $\partial T/\partial t$ at a point 3m away from the cooler.

[Ans: $-0.498℃/s$]

题 4.3 空气由冷风机降温至 18℃ 之后送入房间内，冷空气流入房间并以 $0.15℃/s$ 的速度吸收热量，房间中空气的流动速度为 $0.72\,\text{m/s}$。一段时间之后，假设房间内的温度梯度为 $0.9℃/m$，求距离冷风机 3m 处的 $\partial T/\partial t$ 值。

【答：$-0.498℃/s$】

Problem 4.4 For proper functioning, an electronic instrument on board a balloon should not experience temperature change of more than $±0.006\,\text{K/s}$. The atmospheric temperature is given by

题 4.4 为了使热气球上的仪器正常工作，其外界温度变化速率不应超过 $±0.006\,\text{K/s}$。已知大气温度为

$$T = (288 - 6.5 \times 10^{-3}z)(2 - e^{-0.02t})\,\text{K}$$

where z is the height in meter above the ground and t is the time in hour after the sunrise. Determine the

式中，z 是以米为单位的距地面的高度；t 是以小时为单位的日

minimum allowable rate of ascent when the balloon is at the ground at $t=2$h.

[Ans: 1.12 m/s]

Problem 4.5 The flow through a tube has a velocity given by

$$u = u_{max}\left(1 - \frac{r^2}{R^2}\right)$$

where R is the tube radius and u_{max} is the maximum velocity, which occurs at the tube centerline. (a) Find a general expression for the volume flow rate and the average velocity through the tube, (b) compute the volume flow rate if $R=25$mm and $u_{max}=10$m/s, and (c) compute the mass flow rate if $\rho=1000$kg/m^3.

[Ans:(a) $\frac{1}{2}u_{max}\pi R^2$, $\frac{1}{2}u_{max}$;

(b) 0.00982m^2/s;

(c) 9.82kg/s]

Problem 4.6 A two-dimensional velocity field is given by

$$V = (x - y^2)i + (xy + 2y)j$$

in arbitrary units. At $x=2$ and $y=1$, compute (a) the acceleration components a_x and a_y, (b) the velocity component in the direction $\theta=30°$, and (c) the directions of the maximum velocity and the maximum acceleration.

[Ans: (a) 7m/s^2, 17m/s^2; (b) 2.87m/s; (c) $V=4.123$m/s at $75.96°$ from x-axis, $a=18.385$m/s^2 at $157.62°$ from x-axis]

Problem 4.7 The velocity at a fixed point in a flow field is given by $i+2tj$. Fluid particles passing through that point continue to move with the velocity they had at that point. Find (a) the pathline passing through the point at $t=0$, and (b) the pathline of the particle passing through the point at $t=1$.

[Ans: (a) At $t=0$, $V=i$, the path line is along x-direction; (b) $t=1$, $V=i+2j$, the path line is at an angle $\theta=\arctan(2)$ to the x-direction]

【答：(a) $t=0$时，$V=i$，即迹线是沿着x轴方向；(b) $t=1$时，$V=i+2j$，迹线与x轴夹角为$\theta=\arctan(2)$】

Problem 4.8 The velocity field of a flow is given by

题 4.8 一种流动的速度场表达式为

$$V(x,y,z,t)=10x^2i-20yxj+100tk$$

Determine the velocity and the acceleration of a particle at $x=1\text{m}$, $y=2\text{m}$, $z=5\text{m}$, and $t=0.1\text{s}$.

[Ans: $(10i-40j+10k)\text{m/s}$, $(200i+400j+100k)\text{m/s}^2$]

求$x=1\text{m}$、$y=2\text{m}$、$z=5\text{m}$处$t=0.1\text{s}$时流场中质点的速度和加速度。

【答：$(10i-40j+10k)\text{m/s}$，$(200i+400j+100k)\text{m/s}^2$】

Problem 4.9 Oxygen atoms O enter a chamber shown in Figure 4.14 and exits as O_2. The pressure and temperature of the system at the inlet and exit are the same. The inlet port diameter is 10mm and that of the exit is 20mm. If the inlet velocity is 12m/s, obtain the steady state velocity at the exit assuming all O get transferred to O_2. Also, assume that both O and O_2 obey ideal gas equation of state.

[Ans: 1.5m/s]

题 4.9 氧原子O流入如图4.14所示的容器腔室，以氧气O_2流出。入口与出口处的压力和温度都相同。入口和出口的直径分别为10mm和20mm，如果入口流速为12m/s，假设所有O都转换为O_2，求出口处的稳定流速。假设氧原子O与氧气分子O_2都遵循理想气体状态方程。

【答：1.5m/s】

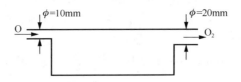

Figure 4.14 Oxygen flow through a chamber

图4.14 流过腔室的氧气流

Problem 4.10 A rotating device to sprinkle water is shown in Figure 4.15 Water enters the rotating device at the center at a rate of $0.03\text{m}^3/\text{s}$ and then it is directed radially through three identical channels of the exit area 0.005m^2 each, perpendicular to the direction of the flow relative to the device. The water leaves at an angle of 30° relative to the device as measured from the radial direction, as shown. If the device rotates

题 4.10 图4.15是一个旋转式淋水器，水由淋水器的中心以$0.03\text{m}^3/\text{s}$的速度流入，然后从三个相同的出口截面为0.005m^2的径向流道流出，与装置相对流动方向垂直。测得流出方向相对于淋水器径向夹角为30°，如图4.15所示。如果淋水

clockwise with a speed of 20 rad/s relative to the ground. Compute the magnitude of the average velocity of the fluid leaving the vane as seen from the ground.

[Ans: 9.16 m/s, at an angle of 79° with respect to the ground (horizontal)]

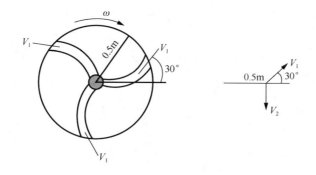

Figure 4.15 Water sprinkler

Problem 4.11   A tank is placed on an elevator which starts moving upwards at the time $t=0$ with a constant acceleration a. A stationary hose discharges water into the tank at a constant rate as shown in Figure 4.16. Determine the time required to fill the tank if it is empty at $t=0$.

[Ans: $\dfrac{-V_1 \pm \sqrt{V_1^2 + 2\dfrac{A_2}{A_1}aH}}{a}$]

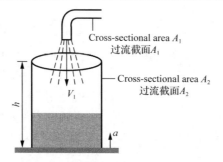

Figure 4.16 Filling of a moving tank

Problem 4.12 Show that the general continuity equation

$$\frac{\partial \rho}{\partial t} + \nabla \cdot (\rho V) = 0$$

can be written in the equivalent form

$$\frac{\mathrm{d}\rho}{\mathrm{d}t} + \rho(\nabla \cdot V) = 0$$

题 4.12 证明通用连续性方程等效形式可以写为

Problem 4.13 Develop the differential form of continuity equation for cylindrical polar coordinates shown by taking an infinitesimal control volume, as shown in Figure 4.17.

题 4.13 通过选取无限小的控制体推导如图 4.17 所示柱坐标系下连续性方程的微分形式。

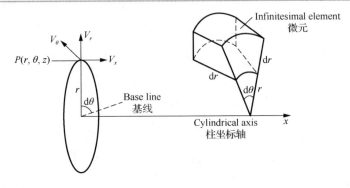

Figure 4.17 Cylindrical polar coordinates

图 4.17 柱坐标系

[Ans: For θ-direction:

$$\rho V_\theta \delta r \delta\theta - \left[(\rho V_\theta \delta r \delta z) + \frac{\partial(\rho V_\theta \delta r \delta\theta)}{\partial\theta}\partial\theta\right]$$

$$= -\frac{1}{2}\frac{\partial \rho V_\theta}{\partial\theta}(r\delta r\delta\theta\delta z)$$

For r-direction:

$$\rho V_r (r\delta\theta\delta z) - \rho V_r (r\delta\theta\delta z) - \rho V_r (r\delta\theta\delta z)$$

$$= -\frac{1}{r}\frac{\partial}{\partial r}(\rho r V_r)(r\mathrm{d}\theta \mathrm{d}z\mathrm{d}r)$$

For z-direction:

【答：θ 方向：

$$\rho V_\theta \delta r \delta\theta - \Big[(\rho V_\theta \delta r \delta z)$$

$$+ \frac{\partial(\rho V_\theta \delta r \delta\theta)}{\partial\theta}\partial\theta\Big]$$

$$= -\frac{1}{2}\frac{\partial \rho V_\theta}{\partial\theta}(r\delta r\delta\theta\delta z)$$

r 方向：

$$\rho V_r (r\delta\theta\delta z) - \rho V_r (r\delta\theta\delta z)$$

$$- \rho V_r (r\delta\theta\delta z)$$

$$= -\frac{1}{r}\frac{\partial}{\partial r}(\rho r V_r)(r\mathrm{d}\theta \mathrm{d}z\mathrm{d}r)$$

z 方向：

$$\rho V_z(r\delta\theta\delta r)-\rho V_z(r\delta\theta\delta r)-\frac{\rho V_z}{\partial z}(r\delta\theta\delta r\delta z)$$
$$=-\frac{\partial\rho V_z}{\partial z}(r\mathrm{d}\theta\mathrm{d}z\mathrm{d}r)]$$

$$\rho V_z(r\delta\theta\delta r)-\rho V_z(r\delta\theta\delta r)$$
$$-\frac{\rho V_z}{\partial z}(r\delta\theta\delta r\delta z)$$
$$=-\frac{\partial\rho V_z}{\partial z}(r\mathrm{d}\theta\mathrm{d}z\mathrm{d}r)】$$

Problem 4.14 A flow field is given by $V=3xi+4yj-5tk\,(\mathrm{m/s})$. (a) Find the velocity at the position (10, 6) at $t=3\mathrm{s}$. (b) What is the slope of the streamlines for this flow at $t=0\mathrm{s}$? (c) Determine the equation of the streamlines at $t=0\mathrm{s}$ up to an arbitrary constant. (d) Sketch the streamlines at $t=0\mathrm{s}$.

[Ans: (a) $V=(30i+24j-15k)\mathrm{m/s}$; (b) $4y/3x$; (c) $\ln y=4/3\ln x+4\ln c$, where c is an arbitrary constant; (d) At $t=0$; the streamlines are straight lines at an angle of 53° to the x-axis]

题 4.14 流场为$V=3xi+4yj-5tk\,(\mathrm{m/s})$。（a）求$t=3\mathrm{s}$时，坐标(10, 6)处的速度；（b）求该流场 t=0s 时流线的斜率；（c）求$t=0\mathrm{s}$到任意常数的流线表达式；（d）画出$t=0\mathrm{s}$时的流线。

【答：（a）$V=(30i+24j-15k)\mathrm{m/s}$；（b）$4y/3x$；（c）$\ln y=4/3\ln x+4\ln c$，其中，$c$是任意常数；（d）$t=0$时流线是与 x 轴成 53° 角的直线】

Problem 4.15 For the fully developed two-dimensional flow of water between two impervious flat plates, shown in Figure 4.18, show that $V_y=0$ everywhere.

题 4.15 两个无渗透平板间充分发展的二维流动如图 4.18 所示，证明流场中处处$V_y=0$。

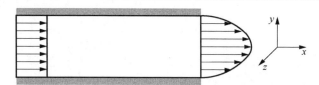

Figure 4.18 Fully developed flow between two plates

图 4.18 两平板之间的充分发展流动

Problem 4.16 Obtain the variation of V_r with r for the radial flow in-between two large discs shown in Figure 4.19. Assume the flow to be incompressible.

[Ans: $V_r=\dfrac{f(z)}{r}$]

题 4.16 图 4.19 所示为两个大圆盘之间的径向流动，假设流动是不可压缩的，求V_r随r变化的表达式。

【答：$V_r=\dfrac{f(z)}{r}$】

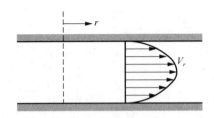

Figure 4.19 Fully developed radially flow between two plates

图 4.19 两个圆盘之间充分发展的径向流动

Problem 4.17 Water enters section 1 at 200m/s and exits at 30° angle at section 2, as shown in Figure 4.20. Section 1 has a laminar velocity profile $u = u_{m1}\left(1 - \dfrac{r^2}{R^2}\right)$, while section 2 has a turbulent profile $u = u_{m1}\left(1 - \dfrac{r}{R}\right)^{1/7}$. If the flow is steady and incompressible (water), find the maximum velocities u_{m1} and u_{m2} in m/s. Assume $u_{av} = 0.5u_m$, for laminar flow, and $u_{av} = 0.82u_m$, for turbulent flow.

[Ans: 5.2m/s, 8.79m/s]

题 **4.17** 如图 4.20 所示，水流以 200m/s 的速度由截面 1 流入，在截面 2 处以 30°角流出。截面 1 处层流速度分布为 $u = u_{m1}\left(1 - \dfrac{r^2}{R^2}\right)$，截面 2 处湍流速度分布为 $u = u_{m1}\left(1 - \dfrac{r}{R}\right)^{1/7}$。如果流动是定常且不可压缩的（水流），求以 m/s 为单位的最大速度 u_{m1} 和 u_{m2}。假设层流 $u_{av} = 0.5u_m$，湍流 $u_{av} = 0.82u_m$。

【答：5.2m/s，8.79m/s】

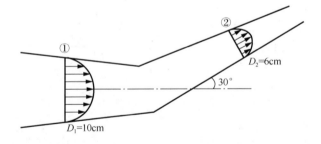

Figure 4.20 Water flow through an elbow

图 4.20 弯头水流

Problem 4.18 A tank with a reentrant orifice called Borda's mouthpiece, is shown in Figure 4.21. The reentrant orifice essentially eliminates flows along the tank walls, so the pressure there is nearly hydrostatic. Water flows out from the mouth as a free jet having an uniform velocity of V_j (m/s). The water jet is

题 **4.18** 如图 4.21 所示，一个容器带有一个称为波达管嘴的内嵌式孔口，这种内嵌式孔口基本上消除了沿容器壁面的流动，因此容器内压力可近似为静压。水从管口以自由射流的形式流出，

surrounded by atmospheric air. Calculate the contraction coefficient, $C_i = A_j / A_o$. Treat the water to be inviscid.

[Ans: 0.5]

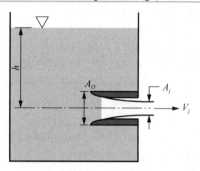

Figure 4.21 Water flow through a Borda mouthpiece

Problem 4.19 Consider a laminar fully developed flow without body forces through a long straight pipe of circular cross-section (Poiseuille flow) shown in Figure 4.22. Apply the momentum equation and show that

$$\tau_{rz} = \frac{p_1 - p_2}{l} \frac{r}{2}$$

Assuming $(p_1 - p_2)/l =$ constant, obtain the velocity profile using the relation.

[Ans: $V_z = \left(\dfrac{p_1 - p_2}{l}\right)\dfrac{1}{4\mu}(R^2 - r^2)$, $\tau_{rz} = -\mu\left(\dfrac{dV_z}{dr}\right)$, $\tau_{rz} = \dfrac{p_1 - p_2}{l} \cdot \dfrac{r}{2}$]

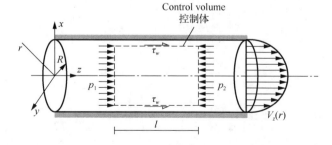

Figure 4.22 Fully developed flow through a pipe

Problem 4.20 Water flows through a circular pipe as shown in Figure 4.23. It enters at A and leaves at C and D. If the velocity at B is $0.8\,\text{m/s}$, and the velocity at C is $2\,\text{m/s}$, calculate the velocities at A and D, and the volumetric flow rate. Assume the flow to be inviscid.

[Ans: $V_A = 1.42\,\text{m/s}, V_D = 13.33\,\text{m/s}, 0.1004\,\text{m}^3/\text{s}$]

题 4.20 如图 4.23 所示圆管水流，由 A 流入，由 C 和 D 流出。如果 B 处速度为 $0.8\,\text{m/s}$，C 处速度为 $2\,\text{m/s}$，计算 A 和 D 处的流速以及体积流量。假设流动是无黏的。

【答：$V_A = 1.42\,\text{m/s}$，$V_D = 13.33\,\text{m/s}$，$0.1004\,\text{m}^3/\text{s}$】

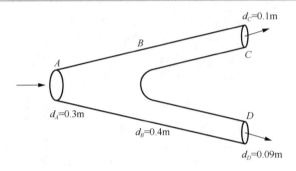

Figure 4.23　Water flow through a pipe with two branches

图 4.23　圆管分支流动

Problem 4.21 Water from a large tank flows through an orifice, as shown in Figure 4.24. If $p_1 = p_{\text{atm}}$, compute the velocity of the jet V when $h = 5\,\text{m}$.

[Ans: $9.9\,\text{m/s}$]

题 4.21 水流从一个巨大水箱的节流口流出，如图 4.24 所示，如果 $p_1 = p_{\text{atm}}$，当 $h = 5\,\text{m}$ 时计算出口射流流速 V。

【答：$9.9\,\text{m/s}$】

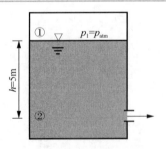

Figure 4.24　Water flow through an orifice

图 4.24　小孔水流

Problem 4.22 In the flow system shown in Figure 4.25, a high-speed water jet issuing from a pipe of area A_j with velocity V_j and the surrounding water flow with

题 4.22 如图 4.25 所示的流动中，高速射流从面积为 A_j 的中间圆管以流速 V_j 流出，与圆管

velocity V_1 get mixed up and the velocity becomes V_2 at section ②. Assuming the velocity profiles to be one-dimension at sections ① and ②, neglecting the viscosity, and assuming the pressure to be uniform across the section ①, using the momentum equation, show that,

$$p_2 - p_1 = \rho \frac{A_j}{A_p}\left(1 - \frac{A_j}{A_p}\right)(V_1 - V_j)^2$$

周围流动速度为V_1的水流混合，在截面②处流速变为V_2。假设截面①、②处流速分布是一维的，忽略黏性，而且截面①处压力均匀，利用动量方程证明

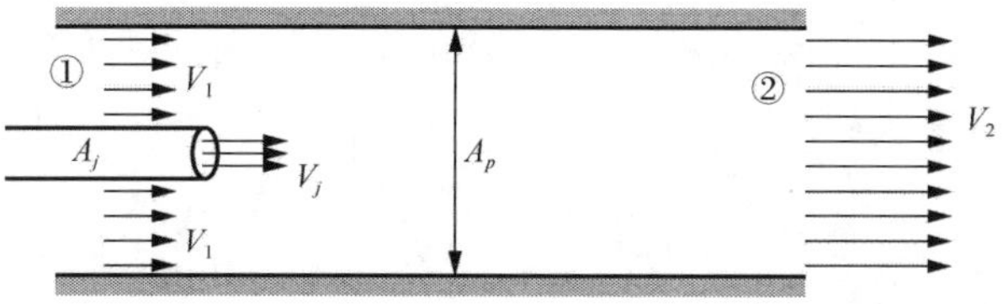

Figure 4.25　Water jet through a pipe with water flow

图4.25　有水流管道中的射流

Problem 4.23　Gasoline with density $\rho = 680 \text{kg/m}^3$ flows from a 30cm diameter pipe in which the pressure is 300kPa into a 15cm diameter pipe in which the pressure is 120kPa. If the pipes are horizontal and viscous effects are negligible, determine the flow rate.

[Ans: $0.420 \text{m}^3/\text{s}$]

题4.23　密度为$\rho = 680 \text{kg/m}^3$的汽油，从一个直径30cm、压力300kPa的管道流入一个直径15cm、压力120kPa的管道。假设管道都是水平的并且忽略黏性作用，求流量。

【答：$0.420 \text{m}^3/\text{s}$】

Problem 4.24　A hang glider soars through the standard sea level air with an air speed of 22m/s. What is the gauge pressure at a stagnation point on the structure?

[Ans: 296.45Pa]

题4.24　一个滑翔机在气流速度为22m/s的标准海平面空气中滑翔，计算滑翔机上驻点处的表压力。

【答：296.45Pa】

Problem 4.25　A two-dimensional fluid motion takes the form of concentric horizontal circular streamlines. Show that, the radial pressure gradient is given by

$$dp/dr = \frac{\rho V_\theta^2}{r}$$

题4.25　具有水平同心圆流线形式的二维流动，证明其径向压力梯度为

where ρ is the density, V_θ is the tangential velocity and r is the radius. Evaluate the pressure gradient for such a flow, defined by $\psi = 2\ln r$, where ψ is the stream function, at a radius of 2m and the fluid density is 10^3kg/m^3.

式中，ρ是密度；V_θ是切向速度；r是半径。由流函数$\psi = 2\ln r$定义的这一流动，当流体密度ρ为10^3kg/m^3时，求半径2m处的压力梯度值。

[Ans: $\dfrac{dp}{dr} = 500\text{N}/\text{m}^3$]　　【答：$\dfrac{dp}{dr} = 500\text{N}/\text{m}^3$】

Problem 4.26　An incompressible flow of a viscous fluid flows parallel to a wide inclined plate in the downhill direction, as shown in Figure 4.26. Treating the flow to be laminar and fully developed, show that the pressure within the fully developed region is a function of y alone. Also, obtain the velocity profile and the volumetric flow rate per unit distance normal to the plane of the flow, $\dot{Q}/w$.

题 4.26　一种不可压缩黏性流体平行于一个宽阔的倾斜平板向下流动，如图 4.26 所示，假设流动是层流并且充分发展。证明充分发展区域压力只是 y 的函数，并推导垂直于流动平面的速度分布和单位平板距离上的体积流量 $\dot{Q}/w$。

[Ans: $V_x = \dfrac{g\sin\theta}{\nu}\left(hy - \dfrac{y^2}{2}\right), \dot{Q}/w = \dfrac{gh^3\sin\theta}{3\nu}$]

【答：$V_x = \dfrac{g\sin\theta}{\nu}\left(hy - \dfrac{y^2}{2}\right),$ $\dot{Q}/w = \dfrac{gh^3\sin\theta}{3\nu}$】

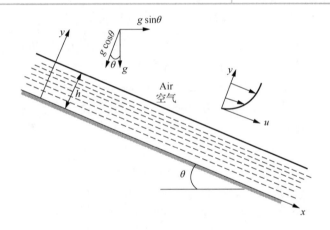

Figure 4.26　Flow over a wide inclined plate

图 4.26　宽阔倾斜平板表面的流动

Problem 4.27　A tank weighs 150N and contains 0.33m³ of water. A force of 220N acts on the tank. What is the value of θ when the free surface of the water assumes a fixed orientation relative to the tank, as shown in Figure 4.27.

[Ans: $-3.71°$]

题 4.27　重 150N 的水罐装有 0.33m³ 的水，受到 220N 的力作用。当自由水面相对于水罐有一个确定方向时，如图 4.27 所示，角 θ 是多少？

【答：$-3.71°$】

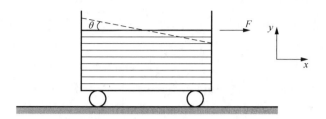

Figure 4.27 A moving tank with water

图 4.27 充满水的移动水罐

Problem 4.28 Consider the pressure-driven flow between stationary parallel plates separated by distance $2h$, shown in Figure 4.28. The velocity field is given by $u = u_{max}[1-(y/h)^2]$, where y is the transverse direction. Evaluate the rates of the linear and angular deformation. Obtain an expression for the vorticity vector ζ. Also, find the location where the vorticity is maximum.

[Ans: Rate of linear deformation zero, Rate of angular deformation in the xy-plane is $-2y\dfrac{u_{max}}{h^2}$, $\zeta = \dfrac{2yu_{max}}{h^2}k$, the vorticity is maximum at $y = \pm h$]

题 **4.28** 间距为 $2h$ 的静止平行平板之间的压差驱动流如图 4.28 所示，速度场可表示为 $u = u_{max}[1-(y/h)^2]$，其中 y 是横向（距离）。求直线变形率和角变形率，以及旋度矢量 ζ 的表达式和最大旋度的位置。

【答：直线变形率为零，xy 平面的角变形率为 $-2y\dfrac{u_{max}}{h^2}$，$\zeta = \dfrac{2yu_{max}}{h^2}k$，最大旋度值在 $y = \pm h$ 处】

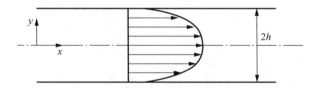

Figure 4.28 Flow between stationary parallel plates

图 4.28 静止平行平板之间的流动

Problem 4.29 Show that, the head loss for laminar, fully developed flow in a straight circular pipe is given by

$$h_l = \dfrac{64}{Re}\dfrac{L}{D}\dfrac{V_{av}^2}{2g}$$

where Re is the Reynolds number defined as $\rho V_{av} d / \mu$.

题 **4.29** 证明直圆管中充分发展的层流水头损失压力损失可表示为

式中，Re 是由 $Re = \rho V_{av} d / \mu$ 定义的雷诺数。

Problem 4.30 The pipe shown in Figure 4.29 contains glycerine at 20℃ flowing at a rate of $8 m^3/h$. Verify that the flow is laminar. Identify

题 **4.30** 如图 4.29 所示的圆管中装有 20℃ 的甘油，其流量为 $8 m^3/h$。证明流动是层流，并判断流动方向是

whether the flow is from right to left or left to right. For glycerine at 20℃, $\mu = 1.49 \text{kg}/(\text{m} \cdot \text{s})$ and $\rho = 1260 \text{kg}/\text{m}^3$.

[Ans: Flow is from right to left, since $\text{head}_B > \text{head}_A$]

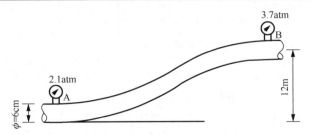

Figure 4.29　Glycerine flow through a pipe

图 4.29　管道中的甘油流动

Problem 4.31　The pipe flow shown in Figure 4.30 is driven by pressurized air in the tank. What gauge pressure p_1 is needed to provide a flow rate $\dot{Q} = 50 \text{m}^3/\text{h}$? Take $\rho = 998 \text{kg}/\text{m}^3$ and $\mu = 0.001 \text{kg}/(\text{m} \cdot \text{s})$.

[Ans: 1890kPa]

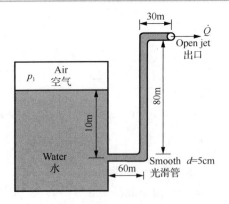

Figure 4.30　Flow through a pipe connected to a tank

图 4.30　容器相连的管道流动

Problem 4.32　A jar with a small circular outlet contains oil on top of water, as shown in Figure 4.31. What is the velocity of this flow at time $t = 0$? What would be the position of the oil-water interface in the jar when the flow stops? The density of the oil is

800kg/m^3.

[Ans: 6.26 m/s, 8 m]

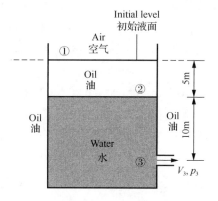

Figure 4.31 A jar with a small circular outlet

Problem 4.33 The water flows into a large tank at a rate of $0.011 \text{m}^3/\text{s}$, as shown in Figure 4.32. The water leaves the tank through 20 holes at the bottom of the tank, each of which produces a stream of 10 mm diameter. Determine the equilibrium height h, for steady state flow.

[Ans: 2.5 m]

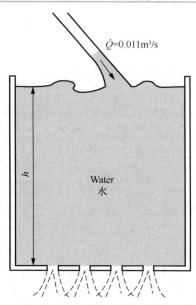

Figure 4.32 Water flow into a large tank with holes at its bottom

Problem 4.34 Consider steady, laminar flow of an incompressible fluid through the horizontal rectangular channel shown in Figure 4.33. Assume that the velocity components in the x and y directions are zero and the only body force is the weight. Starting from Navier-Stokes equations, (a) determine the appropriate set of differential equations and boundary conditions for this problem and (b) show that the pressure distribution is hydrostatic at any given cross-section.

[Ans:(a) $\dfrac{\partial p}{\partial x}=0$, $\dfrac{\partial p}{\partial y}=-\rho g$, $\dfrac{\partial p}{\partial z}=\mu\left(\dfrac{\partial^2 w}{\partial x^2}+\dfrac{\partial^2 w}{\partial y^2}\right)$, the boundary conditions are: at $x=\pm\dfrac{b}{2}$, $w=0$ and at $y=\pm\dfrac{a}{2}$, $w=0$; (b) $p=\rho g y$]

题 4.34 考察流过水平矩形流道的定常不可压缩层流，如图 4.33 所示，假设 x 和 y 方向的速度都为零，受到的体积力只有重力。从纳维-斯托克斯方程出发：（a）推导适用于本流动的一组微分方程及其边界条件；（b）证明在任意给定截面压力都遵循静压分布。

【答：（a） $\dfrac{\partial p}{\partial x}=0$，$\dfrac{\partial p}{\partial y}=-\rho g$，$\dfrac{\partial p}{\partial z}=\mu\left(\dfrac{\partial^2 w}{\partial x^2}+\dfrac{\partial^2 w}{\partial y^2}\right)$，边界条件分别是在 $x=\pm\dfrac{b}{2}$ 处，$w=0$，在 $y=\pm\dfrac{a}{2}$ 处，$w=0$；（b） $p=\rho g y$】

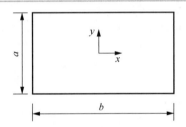

Figure 4.33 Laminar flow through a rectangular channel

图 4.33 矩形流道中的层流

Problem 4.35 A sink of strength $20\text{m}^2/\text{s}$ is situated 3m upstream of a source of strength $40\text{m}^2/\text{s}$, in an uniform horizontal stream. If it is found that there is a point equidistant from both source and sink and 2.5m above the line joining the source and sink, the local velocity is normal to the line joining the source and sink. Find the velocity at this point and the velocity of the undisturbed stream.

[Ans: 0.9373m/s, 2.831m/s]

题 4.35 一个强度为 $20\text{m}^2/\text{s}$ 的点汇放置于强度为 $40\text{m}^2/\text{s}$ 的点源上游3m处，二者均处于一个水平均匀流中。如果能够找到在源汇连线上方2.5m处与源和汇距离相等的一个点，则该点的当地速度垂直于源汇连线。求该点速度大小和未受扰动的来流速度大小。

【答：0.9373m/s，2.831m/s】

Problem 4.36 A jet of water issues from a nozzle with a velocity of 50m/s. If the flow rate is

题 4.36 水射流由喷嘴喷出，速度为 50m/s，如果流量为

$0.22 \text{m}^3/\text{s}$, what is the power of the jet?

[Ans: 275kW]

Problem 4.37 In the boundary layer over the upper surface of an airplane wing, at a point A near the leading edge, the flow velocity just outside the boundary layer is 250km/h. At another point B, which is downstream of A, the velocity outside the boundary layer is 470km/h. If the temperature at A is 288K, calculate the temperature and Mach number at point B.

[Ans: 281.9K, 0.388]

Problem 4.38 A pump discharges $2\text{m}^3/\text{s}$ of water through a pipeline. If the pressure difference between the inlet and outlet of the pump is equivalent to 10m of water, how much power is transmitted to the water from the pump?

[Ans: 196kW]

Problem 4.39 A free vortex flow field is given by $v = \dfrac{q}{2\pi r}$, for $r > 0$. If the flow density $\rho = 1000 \text{kg/m}^3$ and the volume flow rate $q = 20\pi \text{m}^2/\text{s}$, express the radial pressure gradient, $\partial p/\partial r$, as a function of radial distance r, and determine the pressure change between $r_1 = 1\text{m}$ and $r_2 = 2\text{m}$.

[Ans: $\dfrac{100\rho}{r^3}$, 37.5kPa]

Chapter 5　Several Problems of Fluid Dynamics
第 5 章　流体动力学的几个问题

5.1　Introduction

In the previous chapters, we have seen many concepts of fluid flow. With this background, let us observe some of the important aspects of the fluid flow from practical or application point of view in this chapter, for example, the drag forces on bodies due to fluid flow, viscous flows and turbulence flows, which are related to fluid dynamics.

5.2　Viscous Flows

We are familiar with the fact that, the viscosity produces shear force which tends to retard the fluid motion. It works against the inertia force. The ratio of these two forces governs (dictates) many properties of the flow, and the ratio expressed in the form of a non-dimensional parameter is known as the famous Reynolds number, Re_L.

$$Re_L = \frac{\rho \upsilon L}{\mu} \tag{5.1}$$

where υ, ρ are the velocity and density of the flow, respectively, μ is the dynamic viscosity coefficient of the fluid and L is a characteristic dimension. The Reynolds number plays a dominant role in fluid flow analysis. This is one of the fundamental dimensionless parameters which must be matched for similarity considerations in most of the fluid flow analysis. At high Reynolds numbers, the inertia force is predominant compared to viscous forces. At low Reynolds numbers the viscous effects predominate everywhere. Whereas, at high Re the viscous effects confine to a thin region,

5.1　引言

在前述章节中，我们遇到了许多与流动有关的概念。在此基础上，本章我们从实际或应用的角度来观察流动的一些重要特性，例如与流体动力学相关的流体流动对物体产生的阻力、黏性流动和湍流。

5.2　黏性流动

前述我们已经知道流体的黏性会引起阻滞流体运动的剪切力这一事实。黏性力克服惯性力，二者的比值控制（决定）了流动的许多性质。这一比值以一个无量纲参数，也就是雷诺数 Re_L 的形式来表示。

式中，υ、ρ 分别是流体的速度和密度；μ 是流体的动力黏度；L 是特征长度。雷诺数在流动分析中起重要作用，对绝大多数流动分析中的相似条件来说，都是一个基本的必须匹配的无量纲参数。当雷诺数较高时，相对于黏性力而言惯性力起主导性作用；当雷诺数较低时，黏性力处处起主导作用。而在高雷诺数下，黏性效应仅

just adjacent to the surface of the object present in the flow, and this thin layer is termed boundary layer. Since the length and velocity scales are chosen according to a particular flow, when we compare the flow properties at two different Reynolds numbers, only flows with geometric similarity should be considered. In other words, the flow over a circular cylinder should be compared only with the flow past another circular cylinder, whose dimensions can be different but not the shape. The flow in pipes with different velocities and diameters and the flow over aerofoils of the same kind are also some geometrically similar flows. From the above-mentioned similarity consideration, we can infer that the geometric similarity is a prerequisite for dynamic similarity. That is, dynamically similar flows must be geometrically similar, but the converse need not be true. Only similar flows can be compared, that is, when we compare the effect of the viscosity, the changes in the flow pattern due to the body shape should not interfere with the problem.

For calculating Reynolds number, different velocity and length scales are used. Some popular shapes and their length scales we often encounter in fluid flow studies are given in Table 5.1. In the description of Reynolds number here, the quantities with subscript ∞ are at the freestream and quantities without subscript are the local properties. Reynolds number is basically a similarity parameter. It is used to determine the laminar and turbulent nature of flow. Below certain Reynolds number the entire flow is laminar and any disturbance introduced into the flow will be dissipated out by viscosity. The limiting Reynolds number below which the entire flow is laminar is termed lower critical Reynolds number.

局限在一个很薄的紧贴物体壁面的区域之内，这一薄层被称为边界层。由于雷诺数中的长度尺度和速度尺度取决于具体的流动，因此比较两个不同雷诺数下的流动特性时，只能比较几何相似的流动。换言之，圆柱绕流只能与另一个圆柱绕流进行比较，二者的尺寸大小可以不同，但形状必须相同。不同管径和不同流速下的圆管内流动是几何相似的，相同类型的翼型绕流也是几何相似的流动。根据上述相似条件，我们可以推断出，几何相似是动力相似的前提，即动力相似的流动一定是几何相似的，但反之不一定成立。只有彼此相似的流动可以被比较，例如在比较黏性力的作用时，不能引入物体形状因素引起的流动形式变化，否则问题就无法求解。

为计算雷诺数，需要用到不同的速度尺度和长度尺度。表 5.1 给出了流动分析中一些常见的形状和长度尺度。这里在描述雷诺数时，带有下标 ∞ 的量表示无穷远处来流的物理量，没有下标的量表示当地物理量。雷诺数本质上是一个相似性参数，用来判断流动状态是层流还是湍流。当雷诺数低于一定数值时，流动整体为层流状态，任何引入流动中的扰动都会被黏性耗散掉。流动整体能够保持层流状态的最大雷诺数称为下临界雷诺数。

Table 5.1 Some popular shapes and their characteristic lengths
表 5.1 一些常见形状和它们的特征长度

Shape 形状		Characteristic length 特征长度
Circular cylinder 圆柱绕流	$Re_d = \dfrac{\rho_\infty V_\infty d}{\mu_\infty}$	d is cylinder diameter d 是圆柱直径
Aerofoil 翼型	$Re_c = \dfrac{\rho_\infty V_\infty c}{\mu_\infty}$	c is aerofoil chord c 是翼型弦长
Pipe flow (fully developed) 管道流动（充分发展）	$Re_d = \dfrac{\rho \bar{V} d}{\mu}$	$\bar{V}$ is the average velocity $\bar{V}$ 是平均流速 d is the pipe diameter d 是管径
Channel flow (two-dimensional and fully developed) 渠流（二维且充分发展）	$Re_h = \dfrac{\rho \bar{V} h}{\mu}$	$\bar{V}$ is the average velocity $\bar{V}$ 是平均流速 h is the height of the channel h 是渠高
Flow over a grid 栅格流	$Re_m = \dfrac{\rho \upsilon m}{\mu}$	υ is the velocity upstream or downstream of the grid υ 是栅格上游或者下游的流速 m is the mesh size m 是栅格尺寸
Boundary layer 边界层流动	$Re_\delta = \dfrac{\rho \upsilon \delta}{\mu}$	υ is the outer velocity υ 是外部来流速度 δ is the boundary layer thickness δ 是边界层厚度
	$Re_\theta = \dfrac{\rho \upsilon \theta}{\mu}$	θ is the momentum thickness θ 是动量厚度
Flat plate 平板流	$Re_x = \dfrac{\rho \upsilon x}{\mu}$	x is the distance from the leading edge x 是到前缘的距离

Some of the well-known critical Reynolds number are listed below:

Circular cylinder: $Re_w = 200$ (turbulent wake), based on wake width w and wake defect.

Circular cylinder: $Re_d = 1.66 \times 10^5$, based on cylinder diameter d.

Pipe flow: $Re_d = 2300$, based on mean velocity and diameter d.

Channel flow: $Re_h = 1000$ (two-dimensional), based on height h and mean velocity.

Boundary layer: $Re_\theta = 350$, based on freestream velocity and momentum thickness θ.

以下为一些众所周知的临界雷诺数：

圆柱绕流：$Re_w = 200$（湍流尾迹），基于尾迹宽度 w 和尾流耗散长度。

圆柱绕流：$Re_d = 1.66 \times 10^5$，基于圆柱直径 d。

管道流动：$Re_d = 2300$，基于平均流速和管径 d。

渠流：$Re_h = 1000$（二维），基于水深 h 和平均流速。

边界层流动：$Re_\theta = 350$，基于来流速度和动量厚度 θ。

Flat plate: $Re_x = 5 \times 10^5$, based on length x from the leading edge.

It is essential to note that, the transition from laminar to turbulent nature does not take place at a particular Reynolds number but over a range of Reynolds number, because any transition is gradual and not sudden. Therefore, incorporating this aspect, we can define the lower and upper critical Reynolds numbers as follows.

(1) Lower critical Reynolds number is that Reynolds number below which the entire flow is laminar.

(2) Upper critical Reynolds number is that Reynolds number above which the entire flow is turbulent.

(3) Critical Reynolds number is that at which the flow field is a mixture of laminar and turbulent flows.

Note: When the Reynolds number is low due to large viscosity μ the flow is termed stratified flow, for example, flows of tar, honey etc. are stratified flows. When the Reynolds number is low because of the low density, the flow is termed the rarefied flow. For instance, flows in space and very high altitudes, in earth's atmosphere, are rarefied flows.

5.3 Drag of Bodies

When a body moves in a fluid, it experiences forces and moments due to the relative motion of the flow taking place around it. If the body has an arbitrary shape and orientation, the flow will exert forces and moments about all the three coordinate axes, as shown in Figure 5.1. The force on the body along the flow direction is called the drag.

The drag is essentially a force opposing the motion of the body. The viscosity is responsible for a part of the drag force, and the body shape generally determines the overall drag. The drag caused by the viscous effect is termed the frictional drag or skin friction. In the design of transport vehicles, shapes experiencing minimum drag are considered to keep the power consumption at a minimum. Low drag shapes are called streamlined bodies and high drag shapes are termed bluff bodies.

阻力本质上是一个阻止物体运动的力，阻力的大小一部分取决于流体的黏性，但整体通常取决于物体的形状。由黏性引起的阻力叫摩擦阻力或表面阻力。车辆的设计往往采用阻力尽可能小的外形，从而使能耗最低。阻力小的外形称为流线型，相反阻力大的外形称为钝体。

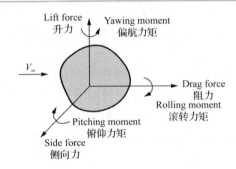

Figure 5.1 Forces acting on an arbitrary body

图 5.1 任意形状物体的受力

The drag arises due to ① the difference in pressure between the front and back regions and ② the friction between the body surface and the fluid. The drag force caused by the pressure imbalance is known as the pressure drag, and the drag due to friction is known as skin friction drag or shear drag. For streamlined body the major portion of the total drag is the skin friction drag, and for a bluff body the major portion of the total drag is the pressure drag.

阻力来自于：①物体前后区域的压力差；②物体表面和流体之间的摩擦。由压差引起的阻力被称为压差阻力，由摩擦引起的阻力被称为表面摩擦阻力或者剪切阻力。一个流线型物体所受的阻力绝大部分为表面摩擦阻力，而一个钝体所受的阻力绝大部分为压差阻力。

5.3.1 Pressure Drag

The pressure drag arises due to the separation of boundary layer, caused by adverse pressure gradient. The phenomenon of the separation, and how it causes the pressure drag can be explained by considering the flow around a body, such as a circular cylinder. If the flow is assumed to be potential, there is no viscosity and

5.3.1 压差阻力

压差阻力是由逆压梯度引起的边界层分离产生的。我们可以用绕流，比如圆柱绕流的例子来说明边界层分离现象以及这一现象是如何产生压差阻力的。若流动是有势的，则无

hence no boundary layer. The flow past the cylinder would be as shown in Figure 5.2, without any separation.

The potential flow around a cylinder will be symmetrical about both the horizontal and vertical planes, passing through the center of the cylinder. The pressure distribution over the front and back surfaces would be identical, and the net force along the freestream direction would be zero. That is, there would not be any drag acting on the cylinder. But in the real flow, because of viscosity, a boundary layer is formed over the surface of the cylinder. The flow experiences a favorable pressure gradient from the forward stagnation point S_1 to the topmost point A on the cylinder at $\theta = 90°$, as shown in Figure 5.3.

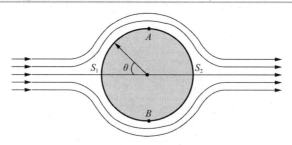

Figure 5.2　Potential flow past a circular cylinder

图 5.2　势流圆柱绕流

Therefore, the flow accelerates from point S_1 to A (that is, from $\theta = 0°$ to $90°$). However, beyond $\theta = 90°$ the flow is subjected to an adverse pressure gradient and hence decelerates. Note that, beyond the topmost point A, the fluid elements find a larger space to relax. Therefore, in accordance with the mass conservation (for subsonic flow), as the flow area increases the flow speed decreases and the pressure increases. Under this condition there is a net pressure force acting against the fluid flow. This process establishes an adverse pressure gradient, leading to the flow separation, as shown in Figure 5.3. In a boundary layer, the velocity near the surface is small, and hence the force due to its momentum is unable to counteract the pressure force. The flow within the boundary layer gets retarded and the velocity near the wall region reduces to zero at some point downstream of A and then the flow is pushed back in the opposite direction, as illustrated in Figure 5.3. This phenomenon is called the flow separation.

边界层中，靠近壁面的流速较低，惯性力不足以克服压力，边界层中流动受阻，接近壁面区域的速度在下游某一点 A 处减为零，然后流动反方向回流，如图 5.3 所示，这一现象称为流动分离。

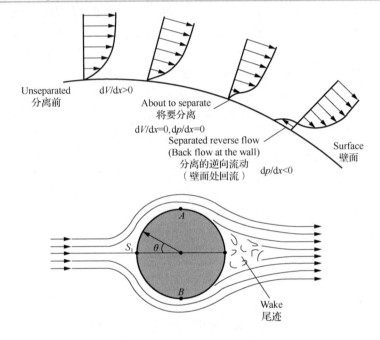

Figure 5.3　Illustration of the separation process

图 5.3　分离过程示意图

The location where the flow leaves the body surface is termed the separation point. For the flow past a cylinder, there are two separation points on either side of the horizontal axis through the center of the cylinder. The separated flow is chaotic and vortex dominated. The separated flow behind an object is also referred to as the wake. Depending on the Reynolds number level, the wake may be laminar or turbulent. An important characteristic of the separated flow is that, it is always unsymmetrical, even for the laminar separation. This is because of the vortices prevailing in the separated zone. As we know, for every vortex there is a specific

流动离开物体表面的位置称为分离点。对于圆柱绕流，在圆柱水平中心轴线两侧存在两个分离点。分离后的流动是以杂乱无章的旋涡为主的流动。在物体后分离的流动也被称为尾迹流。尾迹流可以是层流，也可以是湍流，这取决于雷诺数的大小。分离流动的一个重要特征通常是不对称性，即使是层流分离也是不对称的，其原因是在分离区域交替

frequency and amplitude. Therefore, when the vortices formed at the upper and lower separation points of the cylinder are of the same size and leave the cylinder at the same time, the wake must be symmetric. But this kind of formation of vortices of identical size leaving the upper and lower separation points at the same time is possible only when the geometry of the cylinder is perfectly symmetrical and the freestream flow is absolutely unperturbed and symmetrical about the horizontal plane bisecting the cylinder. But in practice it is not possible to meet these stringent requirements of the flow and geometrical symmetry to establish the symmetrical separation. Owing to this practical constraint all separated flows are unsymmetrical. Indeed, the formation of the vortices at the upper and lower separation points itself is unsymmetrical. When one of them, say the upper one, grows faster, the other one is unable to grow at the same rate. Therefore, only after the faster growing vortex reaches a limiting size possible, for the geometry and Reynolds number combination, and leaves the surface, the growth of the vortex at the opposite side picks up. This retards the growth of the new vortex formed at the location where the vortex leaves the surface. Thus, the alternative shedding of vortices from the upper and lower separation points is established. The alternative shedding of vortices makes the wake chaotic.

Across the separated region, the total pressure is nearly a constant and lower than what it would have been if the flow does not separate. The pressure does not recover completely as in the case of the potential flow. Thus, on account of the incomplete recovery of pressure due to the separation, a net drag force opposing the body motion is generated. We can easily see that, the pressure drag will be small if the separation had taken place later, that is, the area over which the pressure unrecovered is small. To minimize pressure drag, the separation point should be as far as possible from the

出现了旋涡。我们知道，每个旋涡的出现都遵循一个特定的频率和振幅。所以当圆柱上下分离点处形成两个大小相同的旋涡并且同时脱落时，尾迹一定是对称的。但是这种情况只有圆柱形状绝对对称，并且来流完全不受任何扰动，上下对称时才可能发生。而实际条件不可能满足这些流动和几何对称的要求，很难形成对称分离。所以由于实际条件限制，分离流都是不对称的。在上下分离点处形成的旋涡本身就是不对称的，当其中一个，例如上方的旋涡增长较快时，另一个旋涡不会以相同的速率增长。因此，只有当增长快的旋涡增长达到极限离开物体表面时（该极限值取决于流场几何结构和雷诺数），另一侧的旋涡才开始增长，这样就延缓了新旋涡在旋涡脱落处的生成。所以，就产生了旋涡从上下两个分离点处交替脱落的现象，从而形成了杂乱的尾迹流。

在整个分离区域中，流体总压几乎保持不变，其值略低于不发生分离的情况，压力不会完全恢复至势流情况时的数值。所以，物体上就会受到一个阻碍物体运动的由分离引起的压力不完全恢复产生的总阻力。显而易见，分离现象发生得越靠后，即压力不恢复区域越小，则压差阻力越小。为减小压差阻力，应该使分离点尽

leading edge or forward stagnation point. This is true for any shape. Streamlined bodies are designed on this basis and the adverse pressure gradient is kept as small as possible, by keeping the curvature very small. At this stage, it is important to realize that the separated region behind an object is vortex-dominated and these vortices cause considerable pressure loss. Thus the total pressure $p_{0,\text{rear}}$ behind the object is significantly lower than the total pressure $p_{0,\text{face}}$ at the face of the object. This difference $(p_{0,\text{rear}} - p_{0,\text{face}})$, termed the pressure loss, is a direct measure of the drag. This drag caused by the pressure loss is called the pressure drag. This is also referred to as form drag, because the form or shape of the moving object dictates the separation and the expanse of the separated zone. The separation zone behind an object is also referred to as the wake. That is, the wake is the separated region behind an object (usually a bluff body) where the pressure loss is severe. It is essential to note that, what is meant by pressure loss is the total pressure loss, and there is nothing like static pressure loss.

The separation of the boundary layer depends not only on the strength of the adverse pressure gradient but also on the nature of the boundary layer, namely, laminar or turbulent. A laminar flow has tendency to separate earlier than a turbulent flow. This is because the laminar velocity profiles in a boundary layer have lesser momentum near the wall. This is conspicuous in the case of flow over a circular cylinder. Laminar boundary layer separates nearly at $\theta = 90°$ whereas, for a highly turbulent boundary layer the separation is delayed and the attached flow continues up to as far as $\theta = 150°$ on the cylinder. The reduction of pressure drag when the boundary layer changes from the laminar to the turbulent is of

量远离圆柱前缘或者前驻点，这一结论适用于任何形状。基于这一结论，人们把物体设计成流线型，从而通过采用小曲率壁面使逆压梯度被尽可能抑制。这里尤其应意识到的是，物体后面的分离区域是旋涡主导的，而且这些旋涡引起较大的压力损失。所以物体后面的总压 $p_{0,\text{rear}}$ 比前缘处的总压 $p_{0,\text{face}}$ 低很多，这个被称为压力损失的差值 $(p_{0,\text{rear}} - p_{0,\text{face}})$ 直接反映阻力的大小。因此，这一由压力损失引起的阻力也称为压差阻力。由于运动物体的形态和形状决定了流动分离和分离区域的扩展，这一阻力也称为型阻，物体后的分离区域也被称为尾迹，即尾迹是指物体（通常为钝体）之后压力损失较大的分离区域。要注意，这里所说的压力损失是总压损失，没有静压损失这一概念。

边界层的分离不仅与逆压梯度的强度有关，也与边界层的本质有关，即边界层是层流边界层还是湍流边界层。因为层流边界层速度分布在近壁面处，具有更小的惯性，层流边界层一般比湍流边界层更早分离。圆柱绕流中二者的区别就很明显。层流边界层在 $\theta = 90°$ 附近发生分离，而对高度湍流的边界层而言，分离延迟至圆柱上大约 $\theta = 150°$ 处。对钝体而言，由层流变为湍流时，压差阻力减小 5 个数量级。分离区

the order of 5 times for bluff bodies. The flow behind a separated region is called the wake. For the low drag, the wake width should be small.

Although separation is shown to take place at well-defined locations on the body, in the illustration in Figure 5.3, it actually takes places over a zone on the surface which can not be identified easily. Therefore, theoretical estimation of separation especially for a turbulent boundary layer is difficult and hence the pressure drag cannot be easily calculated. Some approximate methods exist but they can serve only as guide lines for the estimation of pressure drag.

At this stage, we may wonder about the level of static pressure in the separated flow region or the wake of a body. The total pressure in the wake is found to be lower than that in the freestream, because of the pressure loss caused by the vortices in the wake. But the static pressure in the wake is almost equal to the freestream level. But it is essential to realize that just after the separation, the flow is chaotic and the streaklines do not exhibit any defined pattern. Therefore, the static pressure does not show any specific mean value in the near-wake region and keeps fluctuating. However, beyond some distance behind the object, the wake stabilizes to an extent to assume almost constant static pressure across its width. This distance is about 6 times the diameter for a circular cylinder. Thus, beyond 6 diameter distance the static pressure in the wake is equal to the freestream value.

Note: It may be useful to recall, what is meant by the pressure loss is the total pressure loss and there is nothing like static pressure loss.

域之后流动称为尾迹，为了减小压差阻力，尾迹宽度应该尽可能小。

尽管在图 5.3 中边界层分离发生在物体上一个给定的位置，但在实际流动中分离一般发生在表面上一个不易确定的区域之内。所以，从理论上计算边界层的分离点尤其是湍流边界层的分离点是十分困难的，因此也很难计算压差阻力。我们也可以采用一些近似的计算方法，但只能作为估算压差阻力的参考。

此时我们也许会想知道物体分离区域或尾迹中静态压力的大小。由于尾迹中旋涡引起的压力损失，尾迹中的总压会低于来流中的总压，但是静态压力几乎和来流保持一致。而必须指出的是，分离点之后流动是混乱的，不呈现特定形状的尾迹线，所以在尾迹初始区域静压力也没有一个特定的均值，而是一直在波动。但在下游远离物体一定距离之后，尾迹流趋于稳定，使整个尾迹宽度内的静压基本保持不变。对于圆柱绕流，这个距离大约为 6 倍圆柱直径。所以 6 倍圆柱直径之外尾迹区域静压力与来流静压力相同。

注意：重申一下，我们所说的压力损失是总压损失，没有静压损失这一概念。

5.3.2 Skin Friction Drag

The friction between the surface of a body and the fluid causes viscous shear stress and this force is known as the skin friction drag. Wall shear stress τ at the surface of a body is given by

$$\tau = \mu \frac{\partial V_x}{\partial y} \tag{5.2}$$

where μ is the dynamic viscosity coefficient and $\partial V_x / \partial y$ is the velocity gradient at the body surface $y = 0$. If the velocity profile in the boundary layer is known, then the shear stress can be calculated.

For streamlined bodies, the separated zone being small, major portion of the drag is because of the skin friction. We saw that, bodies are classified as streamlined body and bluff body based on which is dominant among the drag components. A body for which the skin friction drag is a major portion of the total drag is termed streamlined body. A body for which the pressure (form) drag is the major portion is termed bluff body. Turbulent boundary layer results in more skin friction than a laminar one. Examine the skin friction coefficient c_f variation with Reynolds number, for a flat plate kept at zero angle of attack in an uniform stream, plotted in Figure 5.4.

The characteristic length for Reynolds number is the plate length x, from its leading edge. It can be seen from Figure 5.4 that, the c_f is more for a turbulent flow than laminar flow. The friction coefficient is defined as

$$c_f = \frac{\text{Frictional force（摩擦力）}}{\frac{1}{2} \rho V_\infty^2 S} \tag{5.3}$$

where V_∞, ρ are the freestream velocity and density, respectively, and S is the wetted surface area of the flat plate.

5.3.2 表面摩擦阻力

物体壁面和流体之间的摩擦力产生黏性剪切应力，这一摩擦力就是表面摩擦阻力。物体表面的壁面剪切应力 τ 为

式中，μ 是流体的动力黏度，$\partial V_x / \partial y$ 是物体表面 $y=0$ 处的速度梯度。如果我们知道边界层中的速度分布，就可以计算这一剪切应力。

对流线型物体而言，尾迹区域小，阻力绝大部分来自于摩擦阻力。我们看到，根据阻力分量中哪个占主导，物体可被分为流线型物体和钝体两种。在总阻力中表面摩擦阻力占主导的物体称为流线型物体，压差（型）阻力占主导的物体称为钝体。湍流边界层比层流边界层产生更大的表面摩擦阻力。如图 5.4 所示，对均匀来流攻角为零的平板，观察摩擦阻力系数 c_f 随雷诺数 Re 的变化。

雷诺数中的特征长度 x 为到平板前缘的板长。图 5.4 表明，湍流时 c_f 比层流时要大，摩擦系数 c_f 定义为

式中，V_∞ 和 ρ 分别是来流速度和密度；S 是平板的浸湿表面积。

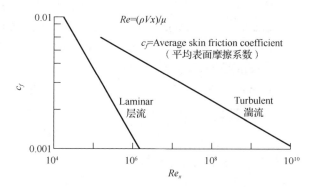

Figure 5.4 Skin fiction coefficient variation with Reynolds number

图 5.4 表面摩擦系数随雷诺数的变化关系

For bluff bodies, the pressure drag is substantially greater than the skin friction drag, and for streamlined bodies the condition is the reverse. In the case of streamlined bodies, such as aerofoil, the designer aims at keeping the skin friction drag as low as possible. Maintaining laminar boundary layer conditions all along the surface is the most suitable arrangement to keep the skin friction low. Though such aerofoils, known as laminar aerofoils, have been designed, they have many limitations. Even a small surface roughness or disturbance can make the flow turbulent, and spoil the purpose of maintaining the laminar flow over the entire aerofoil. In addition, for laminar aerofoils there is a tendency for the flow to separate even at small angles of attack, which severely restricts the use of such aerofoils.

对钝体而言，压差阻力比表面摩擦阻力大很多，而对流线型物体而言，情况则相反。对于流线型物体，例如翼型，设计者要尽可能把表面摩擦阻力降低，其中使整个翼型表面的边界层都处于层流状态是降低表面摩擦的最佳措施。尽管这种层流翼型已经被设计出来，但它们存在很多局限性。即使翼型表面很小的粗糙不平或者来流中很小的扰动都可能使流动从层流变为湍流，从而无法达到在整个翼型表面保持层流状态的目的。此外，即使在小攻角运动时，层流翼型的边界层也有发生分离的趋势，这也严重限制了层流翼型的使用。

5.3.3 Comparison of Drag of Various Bodies

In low-speed flow past geometrically similar bodies with identical orientation and relative roughness, the drag coefficient should be a function of the Reynolds number only.

5.3.3 不同物体阻力的比较

低速来流绕过具有相同方向和相同相对表面粗糙度的几何相似物体时，其阻力系数应该仅为雷诺数的函数

$$C_D = f(Re) \tag{5.4}$$

The Reynolds number is based upon freestream velocity V_∞ and a characteristic length L of the body. The drag coefficient C_D could be based upon L^2, but it is customary to use a characteristic area S of the body instead of L^2. Thus, the drag coefficient becomes

雷诺数用来流速度 V_∞ 和物体的特征长度 L 来计算。阻力系数 C_D 或许可以用 L^2 来计算，但是一般会采用特征面积 S 来代替 L^2，即

$$C_D = \frac{\text{Drag（阻力）}}{\frac{1}{2}\rho V_\infty^2 S} \tag{5.5}$$

(The factor $\frac{1}{2}$, in the denominator of the C_D expression, is our traditional tribute to Euler and Bernoulli.) The area S is usually one of the following three types.

（C_D 表达式分母中的系数 $\frac{1}{2}$ 是我们根据传统为致敬 Euler 和 Bernoulli 而添加在式中的。不加 $\frac{1}{2}$ 该式也成立。）面积 S 一般采用以下三种形式之一。

(1) Frontal area of the body as seen from the flow stream. This is suitable for thick stubby bodies, such as spheres, cylinders, cars, missiles, projectiles, and torpedos.

(2) Planform area of the body as seen from above. This is suitable for wide flat bodies such as aircraft wings and hydrofoils.

(3) Wetted area. This is appropriate for surface ships and barges.

While we use drag or other fluid (aerodynamic) force data, it is important to note what length and area are being used to scale the measured coefficients.

Table 5.2 gives a few data on drag, based on the frontal area, of two dimensional bodies of various cross-sections, at $Re \geq 10^4$.

Drag coefficients of sharp-edged bodies, which has a tendency to experience flow separation regardless of the nature of boundary layer, are insensitive to Reynolds number. The elliptic cylinders, being smoothly rounded, have the laminar turbulent transition effect and are therefore quite sensitive to the nature of the boundary layer (that is, laminar or turbulent).

（1）从来流方向观察的物体迎风面积，适用于短而粗的物体，例如球体、圆柱、汽车、导弹、炮弹和鱼雷。

（2）从上往下看的物体平面面积，适用于扁平状的物体，如机翼和水翼。

（3）浸湿面积，适用于舰艇和船舶的表面。

当使用阻力或者其他液流（气动）力数据时，尤其要注意使用的是针对哪个长度和面积来进行测量的系数。

表 5.2 给出了一些 $Re \geq 10^4$ 的不同形状二维物体迎风面积阻力系数。

对于锐边物体，无论其边界层是哪种边界层，都有分离的趋势，其阻力系数都对雷诺数不敏感。对于光滑的椭圆柱体，具有层流湍流转捩效应，因此其阻力系数对边界层的本质（即层流边界层还是湍流边

Table 5.3 lists drag coefficients of some three-dimensional bodies. For these bodies also we can conclude that sharp edges always cause flow separation and high drag which is insensitive to Reynolds number.

Rounded bodies, such as ellipsoid, have drag which depends upon the point of separation, so that both Reynolds number and the nature of boundary layer are important. Increase of body length will generally decrease the pressure drag by making the body relatively more slender, but sooner or later the skin friction drag will catch up. For a flat-faced cylinder, the pressure drag decreases with L/d but the skin friction drag increases, so that minimum drag occurs at about $L/d=2$.

界层）十分敏感。

表 5.3 列出了一些三维物体的阻力系数。对这些物体，我们也可以得出结论，锐边总是会引起流动分离，并产生较大的阻力，因此阻力系数对雷诺数不敏感。

圆形物体例如椭圆体，其阻力取决于分离点的位置，所以雷诺数和边界层属性都起重要作用。通过使物体变得更加细长来增加物体长度，这样通常可以减小压差阻力，但是随着长度的增加，表面摩擦阻力终究也会增大。对前缘为平面的圆柱而言，压差阻力随着 L/d 减小而减小，但是表面摩擦阻力会增大，所以在大约 $L/d=2$ 时阻力最小。

Table 5.2 Drag of two-dimensional bodies at $Re \geqslant 10^4$[1]

表 5.2 $Re \geqslant 10^4$ 时不同形状二维物体的阻力系数[1]

Shape 形状	C_D
○ ---	1.17+
(-----	1.20
◐ ---	1.16
▯ ---	1.60
◇	1.55
◁ ---	1.55
Vortex street 涡街	1.98
▷ ---	2.00
)	2.30

Shape 形状	C_D
(wedge shape)	2.20
(square block)	2.05+

Note: + In subcritical flow
注：+在底层流动中

Table 5.3　Drag of three-dimensional bodies at $Re \geqslant 10^{4}$[1]
表 5.3　$Re \geqslant 10^{4}$ 时不同形状三维物体的阻力系数[1]

Shape 形状	C_D
String support 张线尾撑 (sphere)	0.47+
(hemisphere, flat side right)	0.38
(hemisphere, curved side right)	0.42
(ellipsoid)	0.59+
Cube 立方体	0.80+
60° (cone)	0.50
Separation 分离 (flat plate)	1.17
(hemisphere, flat side right)	1.17
(hemisphere, curved side right)	1.42
(hemisphere)	1.38
Cube 立方体	1.05+

Note: +Tested on wind tunnel floor
注：+在风洞地板上测得

5.4　Turbulence

Turbulent flow is usually described as the flow with irregular fluctuations. In nature, most of the flows are turbulent. Turbulent flows have characteristics

5.4　湍流

湍流通常被描述为具有不规则脉动的流动状态，自然界中大多数流动都是湍流。湍流

which are appreciably different from those of laminar flows. We have to explain all the characteristics of turbulent flow to completely describe it. Incorporating all the important characteristics, the turbulence may be described as a three-dimensional, random phenomenon, exhibiting multiplicity of scales, possessing vorticity, and showing very high dissipation. Turbulence is described as a three-dimensional phenomenon. This means that, even in a one-dimensional flow field the turbulent fluctuations are always three-dimensional. In other words, the mean flow may be one- or two- or three-dimensional, but the turbulence is always three-dimensional. From the above discussions, it is evident that turbulence can only be described and cannot be defined.

A complete theoretical approach to turbulent flow similar to that of laminar flow is impossible because of the complexity and apparently random nature of the velocity fluctuations in a turbulent flow. Nevertheless, semitheoretical analysis aided by limited experimental data can be carried out for turbulent flows, with instruments which have the capacity to detect high frequency fluctuations. For flows at very low-speeds, say around 20m/s, the frequencies encountered will be 2 to 500Hz. Hot-wire anemometer is well suited for measurements in such flows. A typical hot-wire velocity trace of a turbulent flow is shown in Figure 5.5.

Turbulent fluctuations are random, in amplitude, phase and frequency. If an instrument such as a pitot-static tube, which has a low frequency response of the order of 30 seconds, is used for the measurement of velocity, the manometer will read only a steady value, ignoring the fluctuations. This means that, the turbulent flow consists of a steady velocity component which is independent of time, over which the fluctuations are superimposed, as shown in Figure 5.5(b). That is,

具有明显不同于层流的流动特征，所以要想完全描述湍流我们必须逐一介绍湍流的各个特征。如果将湍流的所有重要特征归纳在一起，湍流可以描述为一种三维的随机现象，同时表现出多尺度性、旋涡性和高耗散性。湍流被描述为一种三维现象，也就意味着即使在一维流场中，湍流脉动也是三维的。换言之，主流可以是一维、二维或者三维的，但是湍流脉动总是三维的。上述讨论说明，湍流只能被描述而不能被定义。

由于湍流中速度脉动的复杂性和明显的随机性，不可能有一个和层流相似的适用于湍流的纯理论方法。然而，借助一些能够测量高频湍流脉动的实验仪器测得的有限实验数据，可以开展湍流的半理论分析。对于极低速流动，比如流速在20m/s左右的流动，湍流脉动频率一般为2～500Hz，热线风速仪可以用来测量这样的流动。图5.5是一个典型的热线风速仪测量的湍流速度曲线。

湍流脉动在幅值、相位和频率上都是随机的。如果用一个测量仪，例如具有30s数量级较低频率响应的皮托管，来测量湍流速度，仪表将只会显示一个稳态的数值，而略掉了湍流脉动值。这意味着湍流包含一个不随时间变化的稳态速度分量，在其上叠加了一个湍流

	脉动，如图 5.5（b）所示，即
$$U(t) = \bar{U} + u'(t) \qquad (5.6)$$	
where $U(t)$ is the instantaneous velocity, $\bar{U}$ is the time averaged velocity, and $u'(t)$ is the turbulent fluctuation around the mean velocity. Since $\bar{U}$ is independent of time, the time average of $u'(t)$ should be equal to zero. That is,	式中，$U(t)$ 是瞬时速度；$\bar{U}$ 是平均速度；$u'(t)$ 是平均速度上的湍流脉动。因为 $\bar{U}$ 与时间无关，假设时间 t 足够大，$u'(t)$ 的时均值应该等于零，即
$$\frac{1}{t}\int_0^t u'(t) = 0, \quad \overline{u'} = 0$$	
provided the time t is sufficiently large. In most of the laboratory flows, averaging over a few seconds is sufficient if the main flow is kept steady.	对大多数实验室中的流动，如果主流保持稳定，则平均时间选几秒钟就足够了。

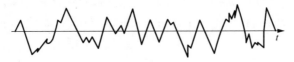

（a）Typical turbulent velocity trace
（a）典型的湍流速度轨迹

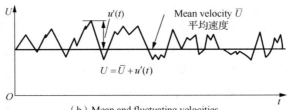

（b）Mean and fluctuating velocities
（b）平均速度和脉动速度

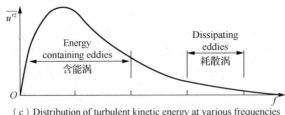

（c）Distribution of turbulent kinetic energy at various frequencies
（c）不同频率下湍流动能的分布

Figure 5.5 Hot-wire trace of a turbulent flow

图 5.5 湍流的热线风速仪曲线

In the beginning of this section, we saw that the turbulence is always three-dimensional in nature even if the main flow is one-dimensional. For example, in a fully developed pipe or channel flow, as for as the	在这一节的开头，我们看到，即使主流是一维运动，湍流本质上也总是三维的。例如充分发展的管流或者渠流中，考虑平

mean velocity is concerned only the *x*-component of velocity $\bar{U}$ alone exists, whereas all the three components of turbulent fluctuations u', v' and w' are always present. The intensity of the turbulent velocity fluctuations is expressed in the form of its root mean square value. That is, the velocity fluctuations are instantaneously squared, then averaged over certain period and finally square root is taken. The root mean square (RMS) value is useful in estimating the kinetic energy of fluctuations. The turbulence level for any given flow with a mean velocity $\bar{U}$ can also be expressed as a turbulence number, *n*, defined as

$$n = 100 \frac{\sqrt{u'^2 + v'^2 + w'^2}}{3\bar{U}} \tag{5.7}$$

In the laboratory, turbulence can be generated in many ways. A wiremesh placed across an air stream produces turbulence. This turbulence is known as grid turbulence. If the incoming air stream as well as the mesh size is uniform then the turbulent fluctuations behind the grid are isotropic in nature, that is, u', v', w' are equal in magnitude. In addition to this, the mean velocity is the same across any cross-section perpendicular to the flow direction, that is, no shear stress exists. As the flow moves downstream the fluctuations die—down due to viscous effects. Turbulence is produced in jets and wakes also. The mean velocity in these flows varies and they are known as free shear flows. Fluctuations exist up to some distance and then slowly decay. Another type of turbulent flow often encountered in practice is the turbulent boundary layer. It is a shear flow with zero velocity at the wall. These flows maintain the turbulence level even at large distance, unlike the grid or free shear flows. In wall shear flows or boundary layer type flows, turbulence is produced periodically to counteract the decay.

A turbulent flow may be visualized as a flow made up of eddies of various sizes. Large eddies are first formed, taking energy from the mean flow. They then break up into smaller ones in a sequential manner till they become very small. At this stage the kinetic energy gets dissipated into heat due to the viscosity. Mathematically it is difficult to define an eddy in a precise manner. It represents, in a way, the frequencies involved in the fluctuations. Large eddy means low-frequency fluctuations and small eddy means high-frequency fluctuations encountered in the flow. The kinetic energy distribution at various frequencies can be represented by an energy spectrum, as shown in Figure 5.5(c).

The problem of turbulence is yet to be solved completely. Different kinds of approach are employed to solve these problems. The well-known method is to write the Navier-Stokes equations for the fluctuating quantities and then average them over a period of time, substituting the following in Navier-Stokes equations, Equation (4.11).

$$V_x = \overline{V}_x + u', V_y = \overline{V}_y + v', V_z = \overline{V}_z + w' \tag{5.8a}$$

$$p = \overline{p} + p' \tag{5.8b}$$

where $\overline{V}_x$, $\overline{V}_y$, $\overline{V}_z$ 和 u'、v'、w' are the mean and fluctuational velocity components along x-, y- and z-directions, respectively, and $\overline{p}$, p', respectively, are the mean and fluctuational components of pressure p. Bar denotes the mean values, that is, time averaged quantities.

Let us now consider the x-momentum equation [Equation (4.11)] for a two-dimensional flow.

$$V_x \frac{\partial V_x}{\partial x} + V_y \frac{\partial V_x}{\partial y} = -\frac{1}{\rho} \frac{\partial p}{\partial x} + \nu \left(\frac{\partial^2 V_x}{\partial x^2} + \frac{\partial^2 V_x}{\partial y^2} \right) \tag{5.9}$$

In Equation (5.9), ν is the kinetic viscosity, given by

$$\nu = \mu/\rho$$

Substituting Equation (5.8) into Equation (5.9), we get	将式（5.8）代入式（5.9）中，得到

$$(\overline{V}_x + u')\frac{\partial(\overline{V}_x + u')}{\partial x} + (\overline{V}_y + v')\frac{\partial(\overline{V}_x + u')}{\partial y}$$
$$= -\frac{1}{\rho}\frac{\partial(\overline{p} + p')}{\partial x} + \nu\frac{\partial^2(\overline{V}_x + u')}{\partial x^2} + \nu\frac{\partial^2(\overline{V}_x + u')}{\partial y^2} \quad (5.10)$$

Expanding Equation (5.10), we obtain	将式（5.10）展开可得

$$\overline{V}_x\frac{\partial \overline{V}_x}{\partial x} + \overline{V}_x\frac{\partial u'}{\partial x} + u'\frac{\partial \overline{V}_x}{\partial x} + u'\frac{\partial u'}{\partial x} + \overline{V}_y\frac{\partial \overline{V}_x}{\partial y} + \overline{V}_y\frac{\partial u'}{\partial y} + v'\frac{\partial \overline{V}_x}{\partial y} + v'\frac{\partial u'}{\partial y}$$
$$= -\frac{1}{\rho}\frac{\partial \overline{p}}{\partial x} - \frac{1}{\rho}\frac{\partial p'}{\partial x} + \nu\frac{\partial^2 \overline{V}_x}{\partial x^2} + \nu\frac{\partial^2 u'}{\partial x^2} + \nu\frac{\partial^2 \overline{V}_x}{\partial y^2} + \nu\frac{\partial^2 u'}{\partial y^2} \quad (5.11)$$

In this equation, time average of the individual fluctuations is zero. But the product or square terms of the fluctuating velocity components are not zero. Taking time average of Equation (5.11), we get	这个方程中单个脉动量的时均值为零，但是脉动速度之间的乘积或者其平方项不为零。对式（5.11）求时间平均可以得到

$$\overline{V}_x\frac{\partial \overline{V}_x}{\partial x} + \overline{u'\frac{\partial u'}{\partial x}} + \overline{V}_y\frac{\partial \overline{V}_x}{\partial y} + \overline{v'\frac{\partial u'}{\partial y}} = -\frac{1}{\rho}\frac{\partial \overline{p}}{\partial x} + \nu\frac{\partial^2 \overline{V}_x}{\partial x^2} + \nu\frac{\partial^2 \overline{V}_x}{\partial y^2} \quad (5.12)$$

Equation (5.12) is slightly different from the laminar Navier-Stokes equation [Equation (5.9)].	式（5.12）与层流纳维-斯托克斯方程［式（5.9）］稍有不同。
The continuity equation for the two-dimensional flow under consideration is	所考察的二维流动的连续性方程为

$$\frac{\partial(\overline{V}_x + u')}{\partial x} + \frac{\partial(\overline{V}_y + v')}{\partial y} = 0$$

This can be expanded to result in	时均化展开之后得

$$\frac{\partial \overline{V}_x}{\partial x} + \frac{\partial \overline{V}_y}{\partial y} = 0, \quad \frac{\partial u'}{\partial x} + \frac{\partial v'}{\partial y} = 0 \quad (5.13)$$

The terms involving turbulent fluctuational velocities u' and v' on the left hand-side of Equation (5.12) can be written as	式（5.12）左边包含湍流脉动速度 u' 和 v' 的项可以写为

$$\overline{u'\frac{\partial u'}{\partial x}} + \overline{v'\frac{\partial u'}{\partial y}} = \frac{\partial}{\partial x}\overline{(u'^2)} - \overline{u'\frac{\partial u'}{\partial x}} + \overline{v'\frac{\partial u'}{\partial y}}$$

Using Equation (5.13) the above equation can be expressed as	利用式（5.13），上式可表示为

$$\overline{u'\frac{\partial u'}{\partial x}} + \overline{v'\frac{\partial u'}{\partial y}} = \frac{\partial}{\partial x}\overline{(u'^2)} + \frac{\partial}{\partial y}\overline{(u'v')} \quad (5.14)$$

Combination of Equations (5.12) and (5.14) results in

合并式(5.12)和式(5.14)得到

$$\rho \overline{V}_x \frac{\partial \overline{V}_x}{\partial x} + \rho \overline{V}_y \frac{\partial \overline{V}_x}{\partial y} = -\frac{\partial \overline{p}}{\partial x} + \frac{\partial}{\partial x}\left(\mu \frac{\partial \overline{V}_x}{\partial y} - \rho \overline{u'^2}\right) + \frac{\partial}{\partial y}\left(\mu \frac{\partial \overline{V}_x}{\partial y} - \rho \overline{u'v'}\right) \quad (5.15)$$

The terms $-\rho \overline{u'^2}$ and $-\rho \overline{u'v'}$ in Equation (5.15) are due to turbulence. They are popularly known as Reynolds or turbulent stresses. For a three-dimensional flow, the turbulent stress terms are $\rho \overline{u'^2}$, $\rho \overline{v'^2}$, $\rho \overline{w'^2}$, $\rho \overline{u'v'}$, $\rho \overline{u'w'}$ and $\rho \overline{v'w'}$. Solution to Equation (5.15) is rather cumbersome. Assumptions like eddy viscosity, mixing length are made to find a solution to this equation.

At this stage, it is important to have proper clarity about the laminar and turbulent flows. The laminar flow may be described as "a well orderly pattern where fluid layers are assumed to slide over one another", that is, in laminar flow the fluid moves in layers, or laminas, one layer gliding over an adjacent layer with interchange of momentum only at molecular level. Any tendencies toward instability and turbulence are damped out by viscous shear forces that resist the relative motion of adjacent fluid layers. In other words, "laminar flow is an orderly flow in which the fluid elements move in an orderly manner such that the transverse exchange of momentum is insignificant" and "turbulent flow is a three-dimensional random phenomenon, exhibiting multiplicity of scales, possessing vorticity, and showing very high dissipation".

Turbulent flow is basically an irregular flow. Turbulent flow has very erratic motion of fluid particles, with a violent transverse exchange of momentum.

The laminar flow, though possessing irregular molecular motions, is macroscopically a well-ordered

式（5.15）中 $-\rho \overline{u'^2}$ 和 $-\rho \overline{u'v'}$ 项由湍流效应产生。它们通常被称为雷诺应力或者湍流应力。对于一个三维流动，湍流应力项分别是 $\rho \overline{u'^2}$、$\rho \overline{v'^2}$、$\rho \overline{w'^2}$、$\rho \overline{u'v'}$、$\rho \overline{u'w'}$ 和 $\rho \overline{v'w'}$。由于式（5.15）很难求解，人们构造涡黏性或者混合长度这样的假设来求解这一方程。

现在关键要对层流和湍流有一个正确的区分。层流可以被描述为一种假定流体层之间相互滑动而形成的有序流动形式，即层流中流体呈层状或片状流动，一层流体层在其相邻层间滑动时，只存在分子级的动量交换。流动中的任何不稳定或湍流倾向都被阻碍相邻层流体相对运动的黏性剪切力所抑制，换言之，"层流是一种有序的运动状态，流体微元在其中以一种有序的形式运动，因此动量的横向交换不显著""湍流是一种三维的随机现象，具有尺度多样性，旋涡性以及高耗散性"。

湍流本质上是一种不规则流动，流体质点的运动很不稳定，有非常剧烈的横向动量交换。

层流内部虽然有不规则的分子运动，但是在宏观尺度

flow. But in the case of turbulent flow, there is the effect of a small but macroscopic fluctuating velocity superimposed on a well-ordered flow. A graph of velocity versus time at a given position in a pipe flow would appear as shown in Figure 5.6(a), for laminar flow, and as shown in Figure 5.6(b), for turbulent flow. In Figure 5.6(b) for turbulent flow, an average velocity denoted as $\bar{V}$ has been indicated. Because this average is constant with time, the flow has been designated as steadiness. An unsteady turbulent flow may prevail when the average velocity changes with time, as shown in Figure 5.6(c).

上，仍然是一种有序流动。但就湍流而言，在有序场流动基础上还附加了一个虽然很小但是宏观上脉动的速度。圆管层流和湍流中某一给定位置速度随时间变化，曲线分别如图 5.6（a）和 5.6（b）所示。图 5.6（b）中给出了用 $\bar{V}$ 表示的平均速度。如果这个平均速度不随时间变化，则湍流被称为定常湍流。当平均速度随时间变化时，会形成如图 5.6（c）所示的非定常湍流。

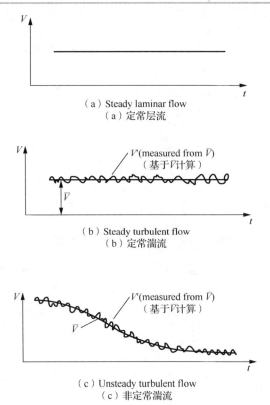

Figure 5.6　Variation of flow velocity with time

图 5.6　流速随时间的变化

5.5　Flow Through Pipes

Fluid flow through pipes with circular and non-circular cross-sections is one of the commonly encountered problems in many practical systems. Flow through pipes is driven mostly by pressure or gravity or both.

Consider the flow in a long duct, shown in Figure 5.7. This flow is constrained by the duct walls. At the inlet, the freestream flow (assumed to be inviscid) converges and enters the tube.

Because of the viscous friction between the fluid and pipe wall, viscous boundary layer grows downstream of the entrance. The boundary layer growth makes the effective area of the pipe to decrease progressively downstream, thereby making the flow along the pipe to accelerate. This process continues up to the point where the boundary layer from the wall grows and meets at the pipe centerline, that is, fills the pipe, as illustrated in Figure 5.7.

5.5　管道流动

流经圆截面或非圆截面管道的流动是许多实际系统中经常遇到的流动问题之一。管道流大多由压力驱动、重力驱动或二者共同驱动。

考察如图 5.7 所示的长管道流动，由于流动受管壁约束，在入口处，来流（假设无黏）收缩进入管道。

由于管壁和流体之间存在黏性摩擦力，所以黏性边界层在入口下游逐渐增长。边界层的增长使得下游的有效过流面积逐渐减小，因此管道中流动逐渐加速，这一过程持续到从壁面增长的边界层在管道中心线处汇合的位置，即边界层充满整个管道为止，如图 5.7 所示。

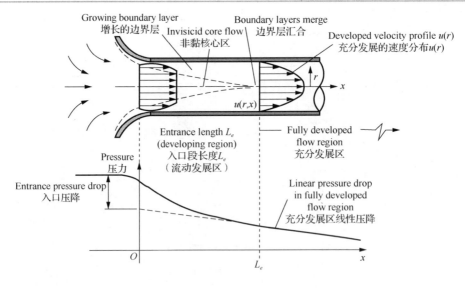

Figure 5.7　Flow development in a long duct

图 5.7　长管中的流动发展

The zone upstream of the boundary layer merging point is called the entrance or flow development length, L_e, and the zone downstream of the merging point is termed fully-developed region. In the fully-developed region, the velocity profile remains unchanged. Dimensional analysis shows that, Reynolds number is the only parameter influencing the entrance length. In the functional form, the entrance length can be expressed as

$$L_e = f(\rho, V, d, \mu)$$

$$\frac{L_e}{d} = f_1\left(\frac{\rho V d}{\mu}\right) = f_1(Re)$$

where ρ, V and μ are the flow density, velocity and viscosity, respectively, and d is the pipe diameter.

For laminar flow, the accepted correlation is

$$\frac{L_e}{d} \approx 0.06 Re_d$$

At the critical Reynolds number $Re_c = 2300$, for pipe flow, $L_e = 138d$, which is the possible maximum development length.

For turbulent flow the boundary layer grows faster, and L_e is given by the approximate relation

$$\frac{L_e}{d} \approx 4.4(Re_d)^{\frac{1}{6}} \tag{5.16}$$

Now, examine the flow through an inclined pipe, shown in Figure 5.8, considering the control volume between sections 1 and 2.

Treating the flow to be incompressible, by volume conservation, we have

$$\dot{Q}_1 = \dot{Q}_2 = \text{constant}（常数）$$

$$V_1 = \frac{\dot{Q}_1}{A_1} = V_2 = \frac{\dot{Q}_2}{A_2}$$

where $\dot{Q}_1$ and $\dot{Q}_2$, respectively, are the volume flow rates and A_1, A_2, V_1 and V_2 are the local areas and velocities, at states 1 and 2. The velocities V_1 and V_2 are equal, since the flow is fully developed and also $A_1 = A_2$.

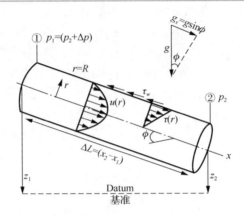

Figure 5.8 Fully-developed flow in an inclined pipe

图 5.8 倾斜管中的充分发展流动

By incompressible Bernoulli's equation, we have 根据不可压缩伯努利方程，有

$$\frac{p_1}{\rho} + \frac{1}{2}V_1^2 + gz_1 = \frac{p_2}{\rho} + \frac{1}{2}V_2^2 + gz_2 \tag{5.17}$$

Since $V_1 = V_2$, we can write from Equation (5.17) the head loss due to friction as 因为 $V_1 = V_2$，从式（5.17）中我们可以写出由摩擦引起的能头损失为

$$h_f = \left(z_1 + \frac{p_1}{\rho g}\right) - \left(z_2 + \frac{p_2}{\rho g}\right) = \Delta z + \frac{\Delta p}{\rho g} \tag{5.18}$$

where $\Delta z = z_1 - z_2$ and $\Delta p = p_1 - p_2$. That is, the head loss (in a pipe), due to friction is equal to the sum of the change in gravity head and pressure head. 式中，$\Delta z = z_1 - z_2$，$\Delta p = p_1 - p_2$。即摩擦引起的能头损失（在管道中）等于重力能头和压力能头变化之和。

By momentum balance, we have 由动量平衡可得

$$\dot{m}(V_1 - V_2) = 0 \tag{5.19}$$

$$\Delta p \pi R^2 + \rho g (\pi R^2)\Delta L \sin\theta - \tau_w (2\pi R)\Delta L = 0$$

$$\Delta p \pi R^2 + \rho g (\pi R^2)\Delta L \sin\theta = \tau_w (2\pi R)\Delta L$$

Dividing throughout by $(\pi R^2)\rho g$, we get 两边同时除以 $(\pi R^2)\rho g$，有

$$\frac{\Delta p}{\rho g} + \Delta L \sin\theta = \frac{2\tau_w}{\rho g}\frac{\Delta L}{R}$$

But $\Delta L \sin\theta = \Delta z$. Thus, 但由于 $\Delta L \sin\theta = \Delta z$，所以

$$\frac{\Delta p}{\rho g} + \Delta z = \frac{2\tau_w}{\rho g}\frac{\Delta L}{R}$$

Using Equation (5.18), we obtain

$$\frac{\Delta p}{\rho g} + \Delta z = h_f = \frac{2\tau_w}{\rho g}\frac{\Delta L}{R} \tag{5.20}$$

In the functional form, the wall shear τ_w may be expressed as

$$\tau_w = F(\rho, V, \mu, d, \varepsilon) \tag{5.21}$$

where μ is viscosity of the fluid, d is the pipe diameter, and ε is the wall roughness height. By dimensional analysis, Equation (5.21) may be expressed as

$$\frac{8\tau_w}{\rho V^2} = f = F\left(Re_d, \frac{\varepsilon}{d}\right) \tag{5.22}$$

where f is called the Darcy friction factor, which is a dimensionless parameter.

Combining Equations (5.20) and (5.22), we obtain the pipe head loss as

$$h_f = f\frac{L}{d}\frac{V^2}{2g} \tag{5.23}$$

This is called the Darcy-Weisbach equation, valid for flow through ducts of any cross-section. Further, in the derivation of the above relation, there was no mention about whether the flow was laminar or turbulent and hence Equation (5.23) is valid for both laminar and turbulent flow. The value of friction factor, f, for any given pipe (that is, for any surface roughness, ε, and diameter, d) at a given Reynolds number can be read from the Moody's diagram (which is a plot of f as a function of Re_d and ε/d), Figure 5.9.

It is essential to note that, in our discussions here it is mentioned that, decrease of pipe area due to boundary layer, results in increase of flow velocity. This is possible only in subsonic flows. When the flow is supersonic, decrease in area will decelerate the flow [2].

利用式（5.18），可得

壁面剪切应力 τ_w 用函数形式可以表示为

式中，μ 是流体黏度；d 是管道直径；ε 是壁面粗糙度。通过量纲分析，式（5.21）可表示为

式中，f 称为达西摩擦因子，一个无量纲参数。

合并式（5.20）和式（5.22），我们可以得到管道的能头损失为

这就称为达西-魏斯巴赫方程，适用于任何截面形状的管道流动。此外，在上述推导中，我们并没有提到流动是层流还是湍流，所以式（5.23）对层流和湍流都适用。对于任何给定的管道（即对任何表面粗糙度 ε 和直径 d），在一个给定雷诺数下，我们可由如图 5.9 所示的穆迪图（f 和 Re_d、ε/d 的函数关系图）读出摩擦因子 f 的数值。

必须要注意，我们的讨论中提到边界层使管路截面积减小，从而导致流速增加，这只适用于亚声速流动。对于超声速流动，过流面积减小反而会使流动减速[2]。

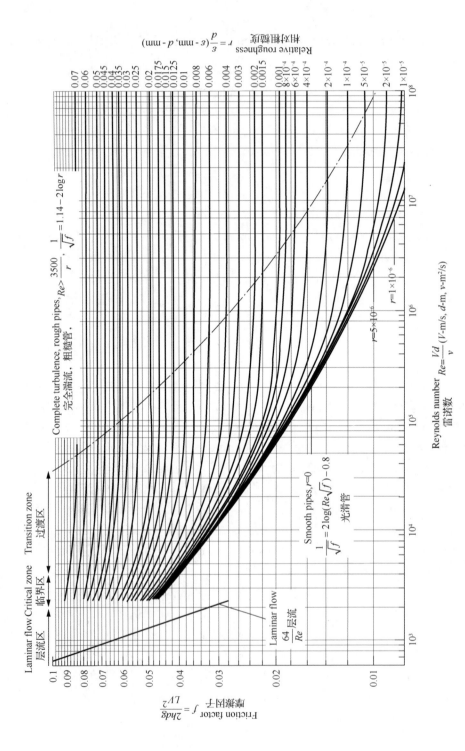

Figure 5.9 Moody's diagram
图5.9 穆迪图

Chapter 5 Several Problems of Fluid Dynamics 流体动力学的几个问题

Even though Moody's diagram is widely used for predicting the value of f applicable for commercial pipes, the concept of an equivalent grain size is open to serious objection. This concept implies that only the height of surface irregularities significantly affects the flow. But in the rough zone of flow, the space of the irregularities is also of great importance. If the irregularities are far apart, the wake of eddies formed by one bump may die away before the fluid encounters the next bump. When the bumps are closer, however, the wake from one bump may interfere with the flow around the next. If the bumps are exceptionally close together the flow may largely skim over the peaks while the eddies may be trapped in the valleys. In Nikuradse's experiments the sand grains were closely packed, and so the spacings may be assumed to be approximately equal to the grain diameter. His results could therefore just be validly taken to demonstrate that f depends on s/d, where s is the average spacing of the grains. The equivalent grain size does not account for the shape of the irregularities. Another factor that may appreciably affect the value of f in large pipes is waviness of the surface, that is, the presence of transverse rides on a large scale than the normal roughness.

尽管穆迪图被广泛应用在市场上购买的管路 f 值的计算中，其中等效粗糙尺寸的概念仍然是有很大异议的，这一概念被认为只有表面不规则高度会严重影响流动，但在流动粗糙区域，不规则间距也是关键。如果两个不规则凸起相隔较远，则由第一个凸起形成的涡流尾迹可能在流到第二个凸起之前就已耗散完毕。但是当两个凸起相隔较近时，第一个凸起引起的尾迹流也会影响到下一个凸起周围的流动。如果两个凸起特别接近，流动可能会直接越过凸起顶部而在凹处形成一个固定的旋涡。在尼库拉泽的实验中，他把沙粒排列得比较紧密，因此可以认为沙粒之间的距离近似等于自身的直径，他的结果刚好可以被用来证明 f 是 s/d 的函数，其中 s 是沙粒之间的平均距离。等效粗糙尺寸并没有考虑不规则颗粒的形状。另一个可能影响大型管道 f 值的因素是表面波度，即大范围的表面横向起伏，而不是通常的粗糙度。

5.6　Flow Past a Circular Cylinder Without Circulation

5.6　无环量的圆柱绕流

A flow pattern equivalent to an irrotational flow over a circular cylinder can be obtained by combining an uniform stream and a doublet with its axis directed against the stream, as shown in Figure 5.10.

圆柱绕流的无旋流动可以等效为一个均匀流和一个轴线与来流方向相反的偶极子的组合，如图 5.10 所示。

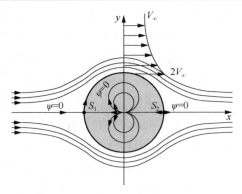

Figure 5.10　Irrotational flow past a circular cylinder without circulation

图 5.10　无环量无旋的圆柱绕流

The points S_1 and S_2 are the stagnation points. The combined stream function becomes	点 S_1 和 S_2 为驻点，叠加之后的流函数为
$$\psi = \psi_{\text{uniform}} + \text{flow}_{\text{doublet}}$$ $$= -\frac{m}{2\pi r}\sin\theta + V_\infty r \sin\theta$$ $$= \left(V_\infty r - \frac{m}{2\pi r}\right)\sin\theta$$	
The potential function for the flow is	流动的势函数为
$$\varphi = \left(V_\infty r + \frac{m}{2\pi r}\right)\cos\theta \qquad (5.24)$$	

It is seen that $\psi = 0$ for all values of θ, showing that the streamline $\psi = 0$ represents a circular cylinder of radius $r = \sqrt{m/(2\pi V_\infty)}$. Let $r = a = \sqrt{m/(2\pi V_\infty)}$. For a given velocity of the uniform flow and a given strength of the doublet, the radius a is constant. Thus, the stream function and potential function of the flow past a cylinder can also be

可见，对于所有的 θ 值，有 $\psi = 0$，表明流线 $\psi = 0$ 代表一个半径 $r = \sqrt{m/(2\pi V_\infty)}$ 的圆柱。令 $r = a = \sqrt{m/(2\pi V_\infty)}$，如果均匀流的速度和偶极子的强度给定，则半径 a 为一个常数。所以该圆柱绕流的流函数和势

expressed as

$$\psi = V_\infty \left(r - \frac{a^2}{r}\right)\sin\theta \tag{5.25}$$

$$\varphi = V_\infty \left(r + \frac{a^2}{r}\right)\cos\theta \tag{5.26}$$

From the flow pattern shown in Figure 5.10, it is evident that the flow inside the circle has no influence on the flow outside the circle. The normal and tangential components of velocity around the cylinder, respectively, are

$$V_r = \frac{\partial \varphi}{\partial r} = V_\infty\left(1 - \frac{a^2}{r^2}\right)\cos\theta$$

$$V_\theta = \frac{1}{r}\frac{\partial \varphi}{\partial r} = -V_\infty\left(1 + \frac{a^2}{r^2}\right)\sin\theta$$

The flow speed around the cylinder is given by

$$|V|_{r=a} = |V_\theta|_{r=a} = 2V_\infty \sin\theta$$

where what is meant by $|V_\theta|$ is the positive value of $\sin\theta$. This shows that there are stagnation points on the surface at $(a, 0)$ and (a, π). The flow velocity reaches a maximum of $2V_\infty$ at the top and bottom of the cylinder, where $\theta = \pi/2$ and $\theta = 3\pi/2$, respectively.

The non-dimensional pressure distribution over the surface of the cylinder is given by

$$C_p = \frac{p - p_\infty}{\frac{1}{2}\rho V_\infty^2} = 1 - \frac{V^2}{V_\infty^2} = 1 - 4\sin^2\theta \tag{5.27}$$

Pressure distribution at the surface of the cylinder is shown by the continuous line in Figure 5.11. The symmetry of the pressure distribution in an irrotational flow implies that, "a steadily moving body experiences no drag". This result, which is not true for actual (viscous) flows where the body experiences drag, is known as d'Alembert's paradox. This discrepancy between the results of inviscid and viscous flows is because of

(1) the existence of tangential stress or skin friction and

(2) drag due to the separation of the flow from the sides of the body and the resulting formation of wake dominated by eddies, in the case of bluff bodies, in the actual flow which is viscous.

The surface pressure in the wake of the cylinder in actual flow is lower than that predicted by irrotational or potential flow theory, resulting in a pressure drag.

是由于：

（1）实际黏性流动中切向应力或者表面摩擦力的存在；

（2）在实际黏性流动中，就钝体绕流而言，物体两侧流动分离而引起的旋涡主导的尾迹流场的出现，使物体受到阻力作用。

实际流动中，尾迹处的圆柱表面压力比由无旋或势流理论计算的压力要低，这就产生了压差阻力。

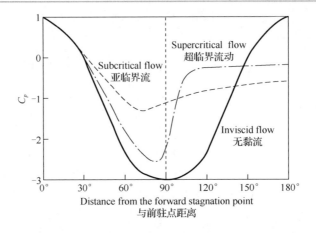

Figure 5.11 Comparison of irrotational and actual pressure (non-dimension) over a circular cylinder

图 5.11 圆柱上各类无旋且真实流动压力（无量纲）分布对比

Note: It is seen that, there are two limits for the C_p, as seen in Figure 5.11. These two limits are $C_p = +1$ and $C_p = -3$, at the forward and rear stagnation points, and the top and bottom locations of the cylinder, respectively. At this stage, it is natural to question about the validity of these limiting values of pressure coefficient for other flows and geometries. Clarifying these doubts is essential from both theoretical and application points of view.

(1) The positive limit of +1 for C_p, at the

注意：在图 5.11 中，C_p 存在两个极值，分别是在前后两个驻点处的 $C_p = +1$ 和上下两个顶点处的 $C_p = -3$ 处。这里我们很自然会提出疑问，这些压力（无量纲）极值是否也适用于其他类型的流动或者物体的其他几何形状，只有同时从理论的和不同应用的角度出发才能解答这一问题。

（1）只要流动是亚声速

forward stagnation point, is valid for all geometries and for both potential and viscous flow, as long as the flow speed is subsonic.

(2) When the flow speed becomes supersonic, there will be a shock ahead of or at the nose of a blunt-nosed and sharp-nosed bodies, respectively. Hence, there are two different speeds at the zones upstream and downstream of the shock. Therefore, the freestream static pressure p_∞ and dynamic pressure $\left(\frac{1}{2}\rho V_\infty^2\right)$ to be used in the C_p relation

$$C_p = \frac{p - p_\infty}{\frac{1}{2}\rho V_\infty^2}$$

have two options, where p is the local static pressure. This makes the C_p at the forward stagnation point sensitive to the freestream static and dynamic pressures, used to calculate it. Therefore, $C_p = +1$ can not be taken as the limiting maximum of C_p, when the flow speed is supersonic.

(3) The limiting minimum of -3, for the C_p over the cylinder in potential flow, is valid only for circular cylinder. The negative value of C_p can take values lower than -3 for other geometries. For example, a cambered aerofoil at an angle of incidence can have C_p as low as -6.

(4) Another important aspect to be noted for viscous flow is that, there is no specific location for rear stagnation point on the body. The flow separates from the body and establishes a wake. The separation is taking place at two locations, above and below the horizontal axis passing through the center of the body. Also, these upper and lower separations are not taking place at fixed points, but oscillate around the separation location, because of vortex formation. Therefore, the negative

的，无论流动是势流还是黏性流动，C_p 在前驻点处存在的最大正值 +1 都适用于所有几何形状的物体。

（2）当流速为超声速时，在钝头和尖头物体头部前缘或者头部分别会形成激波，于是在激波前后会有两种不同的流速，所以 C_p 关系式中所用到的来流静压 p_∞ 和动压 $\left(\frac{1}{2}\rho V_\infty^2\right)$ 两个值有两种不同的选择，

式中，p 是当地静压力。这就使得前驻点处的 C_p 值会随计算所用到的来流静压和动压值的变化而变化，所以，当来流是超声速时，C_p 的最大值不能取 $C_p = +1$。

（3）势流中圆柱上的 C_p 最小值 -3 只适用于圆柱绕流。对于其他形状的物体，C_p 的负值可以比 -3 更低，例如拱形机翼在一定的入射角下 C_p 低值可以达到 -6。

（4）另一方面，对于黏性流动，尤其需要指出的是，物体上的后驻点并没有一个确切的位置，流动从物体上分离之后形成尾迹区。分离一般发生在物体水平中心轴线的上方和下方两个位置。而且上下两个分离点也并不是发生在固定的位置，而是随着旋涡的不断形

pressure at the rear of the body does not assume a specified minimum at any fixed point, as in the case of potential flow. For many combinations of the geometries and flow Reynolds numbers, the negative C_p at the separated zone of the body can assume comparable magnitudes over a large zone of the wake.

5.7 Flow Past a Circular Cylinder With Circulation

We saw that, there is no net force acting on a circular cylinder in a steady irrotational flow without circulation. It can be shown that, a lateral force identical to a lift force on an aerofoil, results when circulation is introduced around the cylinder. When a clockwise line vortex of circulation Γ is superposed around the cylinder in an irrotational flow, the stream function becomes

$$\psi = V_\infty \left(r - \frac{a^2}{r} \right) \sin\theta + \frac{\Gamma}{2\pi} \ln r$$

The tangential velocity component at any point in the flow is

$$V_\theta = -\frac{\partial \psi}{\partial r} = -V_\infty \left(1 + \frac{a^2}{r^2} \right) \sin\theta - \frac{\Gamma}{2\pi r} \quad (5.28)$$

At the surface of the cylinder of radius a, the tangential velocity becomes

$$V_\theta\big|_{r=a} = -2V_\infty \sin\theta - \frac{\Gamma}{2\pi a}$$

At the stagnation point, $V_\theta = 0$, thus,

$$\sin\theta = -\frac{\Gamma}{4\pi a V_\infty} \quad (5.29)$$

For $\Gamma = 0$, the potential flow past the cylinder is symmetrical about both x- and y-directions, as shown in Figure 5.12(a). For this case there is no drag acting on the cylinder.

Chapter 5　Several Problems of Fluid Dynamics　流体动力学的几个问题

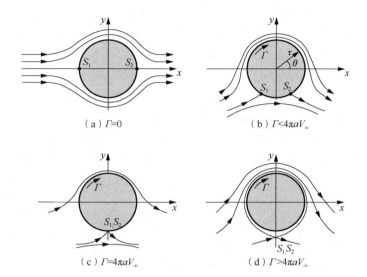

(a) $\Gamma=0$　　(b) $\Gamma<4\pi aV_\infty$

(c) $\Gamma=4\pi aV_\infty$　　(d) $\Gamma>4\pi aV_\infty$

Figure 5.12　Flow past a circular cylinder [(a) without circulation and (b), (c) and (d) with circulation]

图 5.12　圆柱绕流 [(a) 环量为零,(b)、(c)、(d) 环量不为零]

For $\Gamma<4\pi aV_\infty$, two values of θ satisfy Equation (5.29). This implies that there are two stagnation points on the surface, as shown in Figure 5.12(b).	当 $\Gamma<4\pi aV_\infty$ 时,式(5.29)有两个解,这就说明在圆柱表面存在两个驻点,如图 5.12(b) 所示。
When $\Gamma=4\pi aV_\infty$, the stagnation points merge on the negative y-axis, as shown in Figure 5.12(c).	当 $\Gamma=4\pi aV_\infty$ 时,圆柱表面的两个驻点在 y 轴负半轴处重合,如图 5.12(c) 所示。
For $\Gamma>4\pi aV_\infty$ the stagnation points merge and stay outside the cylinder, as shown in Figure 5.12(d). The stagnation points move away from the cylinder surface, since $\sin\theta$ cannot be greater than 1. The radial distance of the stagnation points r for this case can be found from	当 $\Gamma>4\pi aV_\infty$ 时,两个驻点在圆柱外重合,如图 5.12(d) 所示,因为 $\sin\theta$ 不大于 1,所以驻点会离开圆柱表面。此时驻点的径向距离 r 可以由下式计算:
$$V_\theta\big|_{\theta=-\frac{\pi}{2}} = V_\infty\left(1+\frac{a^2}{r^2}\right)-\frac{\Gamma}{2\pi r}=0$$	
This gives	由此可以得出
$$r=\frac{1}{4\pi V_\infty}\left[\Gamma\pm\sqrt{\Gamma^2-(4\pi aV_\infty)^2}\right]$$	
One root of which is $r>a$, and the flow field for this is as shown in Figure 5.12(d), with the stagnation points S_1 and S_2, overlapping and positioned outside	其中一个根满足 $r>a$,此时流场如图 5.12(d) 所示,两个驻点 S_1 和 S_2 在圆柱外重

the cylinder. The second root corresponds to a stagnation point inside the cylinder. But stagnation point for flow past a cylinder cannot be inside the cylinder. Therefore, the second solution is an impossible one.

As shown in Figure 5.12, the location forward and rear stagnation points on the cylinder can be adjusted by controlling the magnitude of the circulation Γ. The circulation which positions the stagnation points in proximity, as shown in Figure 5.12(b) is called subcritical circulation, the circulation which makes the stagnation points to coincide at the surface of the cylinder, as shown in Figure 5.12(c), is called critical circulation, and the circulation which makes the stagnation points to coincide and take a position outside the surface of the cylinder, as shown in Figure 5.12(d), is called supercritical circulation.

To determine the magnitude of the transverse force (vertical to the freestreem flow) acting on the cylinder, it is essential to find the pressure distribution around the cylinder. Since the flow is irrotational, Bernoulli's equation can be applied between a point in the freestream flow and a point on the surface of the cylinder. Bernoulli's equation for incompressible flow is

$$p + \frac{\rho V^2}{2} = p_\infty + \frac{\rho V_\infty^2}{2}$$

Using Equation (5.29), the surface pressure can be found as follows.

At the surface $r = a$, Equation (5.28) gives the local velocity at any point on the surface as

$$V_\theta|_{r=a} = -2V_\infty \sin\theta - \frac{\Gamma}{2\pi a}$$

Substituting this in Bernoulli's equation, we get

$$p_{r=a} + \frac{1}{2}\rho\left(-2V_\infty \sin\theta - \frac{\Gamma}{2\pi a}\right)^2 = p_\infty + \frac{\rho V_\infty^2}{2}$$

that is,

$$p_{r=a} = p_\infty + \frac{1}{2}\rho\left[V_\infty^2 - \left(-2V_\infty\sin\theta - \frac{\Gamma}{2\pi a}\right)^2\right] \tag{5.30}$$

The symmetry of flow about the *y*-axis implies that, the pressure force on the cylinder has no component along the *x*-axis. The pressure force acting in the direction normal to the flow (along *y*-axis) is called the lift force *L* in aerodynamics	流动关于 *y* 轴对称，说明圆柱所受的压力合力在 *x* 轴方向没有分量。在空气动力学中，物体受到的垂直于来流方向（沿着 *y* 轴方向）的压力合力称为升力 *L*。
Consider a cylinder of radius *a* in a uniform flow of velocity V_∞, shown in Figure 5.13.	取一个速度为 V_∞ 的均匀流中半径为 *a* 的圆柱，如图 5.13 所示。

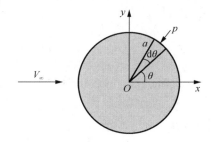

Figure 5.13 Circular cylinder in a uniform flow

图 5.13 均匀流中的圆柱

The lift acting on the cylinder is given by	圆柱受到的升力为
$$L = -\int_0^{2\pi} p_{r=a} a\sin\theta\, d\theta$$	
Substituting Equation (5.30), and integrating we obtain the lift as	将式（5.30）代入上式并积分可以得到升力为
$$L = \rho V_\infty \Gamma \tag{5.31}$$	
where we have used	这里我们用到了
$$\int_0^{2\pi}\sin\theta\, d\theta = \int_0^{2\pi}\sin^3\theta\, d\theta = 0$$	
It can be shown that, Equation (5.31) is valid for irrotational flows around any two-dimensional shape, not just for circular cylinders alone. The expression for lift in Equation (5.31) shows that, the lift force proportional to circulation Γ is of fundamental importance in aerodynamics. Wilhelm Kutta (1902), the German mathematician, and Nikolai Zhukovsky (1906), the Russian aerodynamicist, have proved the relation for lift,	可以证明，式（5.31）不是只适用于圆柱绕流，而是对任意二维形状的无旋绕流都适用。式（5.31）的升力表达式说明，升力与环量的大小成正比，这是一个空气动力学的基本规律。德国数学家威廉·库塔（于1902年）和俄

given by Equation (5.31), independently, and is called the Kutta-Zhukovsky lift theorem (the name Zhukovsky is transliterated as Joukowsky in older texts). The circulation developed by certain two-dimensional shapes, such as aerofoil, when placed in a stream can be explained with vortex theory. It can be shown that, viscosity of the fluid is responsible for the development of circulation. The magnitude of circulation, however, is independent of viscosity, and depends on the flow speed V_∞, the shape and orientation of the body to the freestream direction.

For a circular cylinder in a potential flow, the only way to develop circulation is by rotating it in a flow stream. Although viscous effects are important in this case, the observed pattern for high rotational speeds displays a striking similarity to the ideal flow pattern for $\varGamma > 4\pi a V_\infty$. When the cylinder rotates at low speeds, the retarded flow in the boundary layer is not able to overcome the adverse pressure gradient behind the cylinder. This leads to the separation of the real (actual) flow, unlike the irrotational flow which does not separate. However, even in the presence of separation, observed speeds are higher on the upper surface of the cylinder, implying the existence of a lift force.

A second reason for a rotating cylinder generating lift is the asymmetry to the flow pattern, caused by the delayed separation on the upper surface of the cylinder. The asymmetry results in the generation of the lift force. The contribution of this mechanism is small for two-dimensional objects such as circular cylinder, but it is

罗斯空气动力学家尼古拉·茹科夫斯基（于1906年）曾经分别独立证明了式（5.31）的升力关系，所以该关系又被称为库塔-茹科夫斯基升力定理（旧教材曾把Zhukovsky翻译为 Joukowsky）。二维形状的物体，例如置于气流中的翼型，其环量的形成可以用涡流理论来解释。可见，环量的形成与流体的黏性有关，但是环量的大小却与黏性无关，而取决于来流速度V_∞的大小、物体的形状以及物体与来流方向的夹角。

对于势流中的圆柱，唯一能够产生环量的方法就是在流动中旋转圆柱。尽管在这种情况中黏性的影响起重要作用，但是观察到的圆柱高速旋转流动与环量为$\varGamma > 4\pi a V_\infty$的理想流动形式十分相似。当圆柱旋转速度较低时，边界层中的减速流将不足以克服圆柱后部的逆压梯度，这导致真实（实际）流场中边界层的分离，而在无旋流动中是没有这一现象的。即使存在边界层的分离，也会观察到圆柱上方的流速高于圆柱下方流速，表明圆柱上存在升力作用。

旋转圆柱可以产生升力的第二个原因是流动形式的不对称性，这是由圆柱上表面分离点后移引起的，这种流场上下的不对称产生了升力。这种产生升力的机理对圆柱这

the only mechanism for the side force experienced by spinning three-dimensional objects such as soccer, tennis and golf balls. The lateral force experienced by rotating bodies is called the Magnus effect. The horizontal component of the force on the cylinder, due to the pressure, in general is called drag. For the cylinder, shown in Figure 5.13, the drag is given by

$$D = \int_0^{2\pi} p_{r=a} a \cos\theta \, \mathrm{d}\theta$$

It is interesting to note that, the drag is equal to zero. It is important to realize that, this result is obtained on the assumption that the flow is inviscid. In real (actual or viscous) flows the cylinder will experience a finite drag force acting on it due to viscous friction and flow separation.

5.8 Summary

In this chapter, we mainly talked about the mechanism and calculation of drag forces on bodies due to the fluid flow around the bodies. In addition, we also introduced the concepts of flow states and the parameter to determine them. At the end, we showed the calculation for the flow past a circular cylinder with circulation and without circulation as examples.

5.9 Exercises

Problem 5.1 Determine the net vertical force acting on the circular plate shown in Figure 5.14, if the water spreads radially on it. Neglect the weight of the water on the plate.

[Ans: 79.28kN]

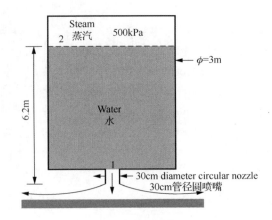

Figure 5.14　Water jet impinging on a horizontal plate

图 5.14　水平板冲击射流

Problem 5.2　The water jet shown in Figure 5.15 strikes normally to a fixed plate. Neglecting gravity and friction, compute the force F required to hold the plate fixed.

[Ans: 502.64 N]

题 5.2　如图 5.15 所示，水流垂直射向固定平板，忽略重力和摩擦力，计算保持平板静止所需的外力 F。

【答：502.64 N】

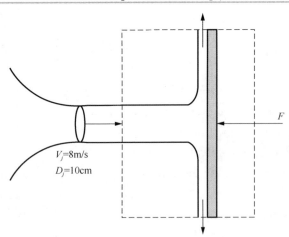

Figure 5.15　Water jet impinging on a vertical plate

图 5.15　垂直平板冲击射流

Problem 5.3　The propulsion device of a boat consists of a pump that takes in water from the river (inlet pipe area A_1) and forces it out as a jet of area A_j, as shown in Figure 5.16. The boat experiences a drag force F when it moves at a speed of V. Show that, the mass flow rate $\dot{m}$ of water through the pump is given by

题 5.3　船的推进装置包括一个从河水中吸水，然后把水喷射出去的泵，泵的进口面积为 A_1，射流面积为 A_j，如图 5.16 所示。当船以速度 V 运动时受到的阻力为 F，证明泵的质量流量 $\dot{m}$ 为

$$\dot{m} = \frac{\rho V A_j}{2} + \sqrt{\frac{\rho^2 V^2 A_j^2}{4} + \rho F A_j}$$

Also, obtain the head developed by the pump. Neglect the hydrostatic pressure forces and viscous losses. [Ans: $\frac{V^2}{2g} \frac{(A_j^2 - A_1^2)}{A_j^2}$]	并推导泵产生的能头大小，忽略静压力和黏性损失。 【答：$\frac{V^2}{2g} \frac{(A_j^2 - A_1^2)}{A_j^2}$】

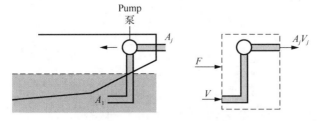

Figure 5.16　Propulsion device of a boat

图 5.16　船的推进装置

Problem 5.4　Water exits to the atmosphere ($p_a = 101\text{kPa}$) through a split nozzle, as shown in Figure 5.17. Duct areas are $A_1 = 0.01\text{m}^2$ and $A_2 = A_3 = 0.005\text{m}^2$. The flow rate $Q_2 = Q_3 = 150\text{m}^3/\text{h}$, and inlet pressure $p_1 = 140\text{kPa}$. Compute the force on the flange bolts at section 1. [Ans: 905.34N]	题 5.4　水由一个分离式喷管排出到大气（$p_a = 101\text{kPa}$）中，如图 5.17 所示。管路截面积 $A_1 = 0.01\text{m}^2$，$A_2 = A_3 = 0.005\text{m}^2$，流量为 $Q_2 = Q_3 = 150\text{m}^3/\text{h}$，入口压力 $p_1 = 140\text{kPa}$。计算截面 1 处法兰螺栓受力。 【答：905.34N】

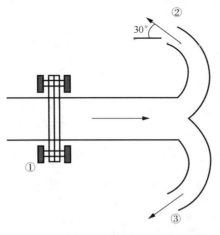

Figure 5.17　Water flow through a split nozzle

图 5.17　分离式喷管水流

Problem 5.5 The fan fixed at the end of a duct (shown in Figure 5.18) sucks air from atmosphere. If the volume displacement of the fan is $1\text{m}^3/\text{s}$, find the power required to run the fan. What is the maximum length h of water that will be sucked up a tube by the flowing air? What is the force required to hold the fan in place?

[Ans: 61W, 0.62cm, 12.25N]

题 5.5 管路末端安装的风扇从大气中吸入空气，如图 5.18 所示。如果风扇的体积流量为 $1\text{m}^3/\text{s}$，计算风扇工作所需的功率以及由流动空气吸入管中的水的最大高度 h，并求使风扇保持不动所需的外力。

【答：61W，0.62cm，12.25N】

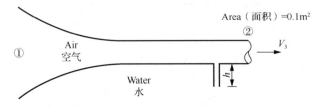

Figure 5.18 A duct with suction fan

图 5.18 抽风机管道

Problem 5.6 For water flow up the sloping channel shown in Figure 5.19, $h_1 = 10\text{mm}$, $V_1 = 3\text{m/s}$ and head $H = 600\text{mm}$. Neglecting losses and assuming uniform flow at 1 and 2, find the downstream depth h_2 and show that, three solutions are possible, of which only one is realistic.

[Ans: -0.272m]

题 5.6 水沿着如图 5.19 所示倾斜流道向上流动，$h_1 = 10\text{mm}$，$V_1 = 3\text{m/s}$，水头 $H = 600\text{mm}$。忽略损失并假设 1 和 2 处都是均匀流动，求下游深度 h_2 并且证明三种可能解中只有一个解是真实的。

【答：-0.272m】

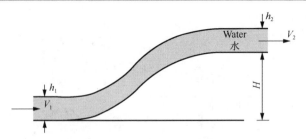

Figure 5.19 Water flow up a sloping channel

图 5.19 倾斜流道中的水流

Problem 5.7 A liquid of density ρ and viscosity μ flows down a stationary wall, under the influence of gravity, forming a thin film of constant thickness h, as shown in Figure 5.20. An up flow of air next to the film

题 5.7 一种密度为 ρ、黏度为 μ 的液体，由于重力的作用，沿着静止墙壁往下流动，形成了厚度恒定为 h 的一层

exerts an upward constant shear stress τ on the surface of the liquid layer, as shown in the figure. The pressure in the film is uniform. Derive expressions for (a) the film velocity V_y as a function of y, ρ, μ, h and τ, and (b) the shear stress τ that would result in a zero net volume flow rate in the film.

[Ans:(a) $V_y = \dfrac{\rho g\left(hx - \dfrac{x^2}{2}\right) - \tau x}{\mu}$; (b) $\tau = \dfrac{2}{3}\rho g h$]

薄膜，如图 5.20 所示。紧挨着薄膜向上流动的空气，在流体层表面施加一个方向向上的恒定剪切应力 τ。薄膜内的压力均匀，求以下各物理量的表达式：（a）用 y、ρ、μ、h 和 τ 表示薄膜速度 V_y；（b）使薄膜中液体总体积流量为零的剪切应力 τ。

【答：（a） $V_y = \dfrac{\rho g\left(hx - \dfrac{x^2}{2}\right) - \tau x}{\mu}$；（b） $\tau = \dfrac{2}{3}\rho g h$】

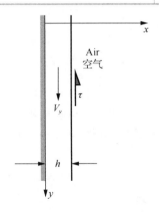

Figure 5.20　Liquid flow over a vertical wall

图 5.20　液体沿垂直壁面的流动

Problem 5.8　For fully developed steady laminar flow of a fluid between two infinite, parallel and stationary flat plates, shown in Figure 5.21, determine (a) the velocity distribution across the channel between the two plates, (b) the maximum and average velocities, and (c) the shearing stress at the wall of the plate and the local frictional coefficient.

[Ans:(a) $V_x = \dfrac{1}{2\mu}\dfrac{\mathrm{d}p}{\mathrm{d}x}(y^2 - h^2)$;

题 **5.8**　两个无限长静止平板之间的充分发展定常层流，如图 5.21 所示，求：（a）两平板之间横跨流道的速度分布；（b）最大速度和平均速度；（c）壁面剪切应力以及当地摩擦系数。

【答：（a） $V_x = \dfrac{1}{2\mu}\dfrac{\mathrm{d}p}{\mathrm{d}x}(y^2 - h^2)$；

(b) $V_{x,\max} = -\dfrac{h^2}{2\mu}\dfrac{dp}{dx}$, $V_{x,av} = \dfrac{2}{3}V_{x,\max}$;
(c) $\tau = \dfrac{3\mu V_{x,av}}{h}$, $C_f = \dfrac{6\mu}{hV_{x,av}}$]

（b）$V_{x,\max} = -\dfrac{h^2}{2\mu}\dfrac{dp}{dx}$，$V_{x,av} = \dfrac{2}{3}V_{x,\max}$；（c）$\tau = \dfrac{3\mu V_{x,av}}{h}$，$C_f = \dfrac{6\mu}{hV_{x,av}}$】

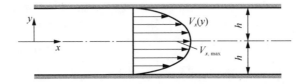

Figure 5.21 Fully developed steady laminar flow between large stationary plates

图 5.21 两个无限长静止平板之间的充分发展定常层流

Problem 5.9 During a study of a certain flow system the following equation relating the pressure p_1 and p_2 at two points was developed.

题 5.9 某一流动系统的研究中，两点之间压力 p_1 和 p_2 可建立起如下关系式：

$$p_2 = p_1 + \frac{flV}{Dg}$$

where V is a velocity, l is the distance between two points, D is a diameter, g is the gravitational acceleration, and f is a dimensionless coefficient. Is the equation dimensionally consistent?

[Ans: No]

式中，V 是速度；l 是两点之间的距离，D 是直径；g 是重力加速度；f 是无量纲系数。该方程是否是量纲一致的。

【答：否】

Problem 5.10 A large movable plate is located between two fixed plates, as shown in Figure 5.22. Two Newtonian fluids having viscosities indicated in the figure are contained between the plates. Determine the magnitude and direction of the shearing stresses that act on the fixed walls when the moving plate has a velocity of 4m/s, as shown. Assume the viscosity distribution between the plates to be linear.

[Ans: 13.33Pa on the upper plate, 13.33Pa on the lower plate, both the stresses act in the direction of the moving plate]

题 5.10 在两个固定平板之间有一个可移动大平板，如图 5.22 所示。平板之间分别充满了两种牛顿流体，其黏性如图中标注。当可移动平板以 4m/s 的速度移动时，求两个固定平板所受的剪切应力大小和方向。假设平板之间的速度分布都是线性的。

【答：上平板13.33Pa，下平板13.33Pa，方向都和平板移动方向一致】

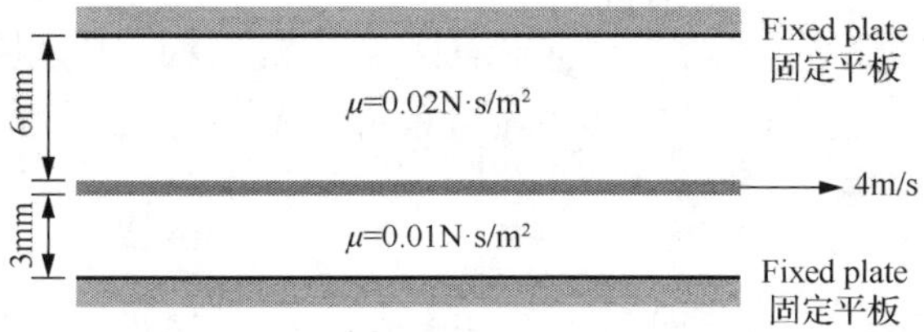

Chapter 5　Several Problems of Fluid Dynamics　流体动力学的几个问题

Figure 5.22　A large plate moving in a fluid contained between two fixed plates

图 5.22　在两个固定平板之间液体中移动的大平板

Problem 5.11　The velocity in a certain flow field is given by the equation

$$V = 3yz^2 i + xzj + yk$$

Determine the expressions for the three rectangular components of acceleration.

[Ans: $a_x = 3xz^3 + 6y^2 z$, $a_y = 3yz^3 + xy$, $a_z = xz$]

题 5.11　某一流场中的速度分布可用方程表示为

求加速度三个直角坐标分量的表达式。

【答：$a_x = 3xz^3 + 6y^2 z$, $a_y = 3yz^3 + xy$, $a_z = xz$】

Problem 5.12　Determine an expression for the vorticity of the flow field described by $V = x^2 yi - xy^2 j$. Is the flow irrotational?

[Ans: $\zeta = \zeta_z k = -(x^2 + y^2)k$. The flow is not irrotational, since the vorticity is not zero]

题 5.12　推导由 $V = x^2 yi - xy^2 j$ 描述的流场旋度表达式，并说明该流场是否无旋。

【答：$\zeta = \zeta_z k = -(x^2 + y^2)k$，流动是有旋的，因为旋度不为零。】

Problem 5.13　For a certain incompressible, two-dimensional flow field the velocity component in the y-direction is given by the equation

$$v = x^2 + 2xy$$

Determine the velocity component in the x-direction so that the continuity equation is satisfied.

[Ans: $u = -x^2 + f(y)$]

题 5.13　一个二维不可压缩流场，其 y 方向速度分量表达式为

求满足连续性方程的 x 方向速度分量。

【答：$u = -x^2 + f(y)$】

Problem 5.14　The velocity potential for a certain flow field is $\varphi = 4xy$. Determine the corresponding stream function.

[Ans: $\psi = 2(y^2 - x^2) + \text{constant}$]

题 5.14　某一流场的速度势函数为 $\varphi = 4xy$，求相应的流函数。

【答：$\psi = 2(y^2 - x^2) + $ 常数】

Problem 5.15　In a certain steady, incompressible, inviscid, two-dimensional flow the x-component of velocity is given by

$$u = x^2 - y$$

题 5.15　一个定常二维无黏不可压缩流动中，x 方向的速度分量为

Will the corresponding pressure gradient in the horizontal direction (x-direction) be a function of only y, of only x, or of both x and y?

[Ans: The pressure gradient in the x-direction is a function of only x]

Problem 5.16 In a certain viscous, incompressible, steady flow field without body forces the velocity components are

$$u = ay - b(cy - y^2)$$
$$v = w = 0$$

where a, b, and c are constants. (a) Using Navier-Stokes equations finds an expression for the pressure gradient in the x-direction. (b) For what combination of the constants a, b, and c will the shearing stress, τ_{xy}, be zero at $y = 0$, where the velocity is zero?

[Ans: (a) $\dfrac{\partial p}{\partial x} = 2\mu b$; (b) $a = bc$]

Problem 5.17 An airfoil of chord 2m is tested in an air stream of velocity 50m/s at the sea level. (a) Determine the Reynolds number. (b) If the same airfoil were attached to an airplane flying at the same speed in a standard atmosphere at an altitude of 3000m, what would be the Reynolds number?

[Ans: (a) 6.85×10^6; (b) 5.37×10^6]

Problem 5.18 An incompressible fluid between two large parallel plates, shown in Figure 5.23, is set to motion by suddenly moving the bottom plate at a constant speed U. The governing differential equation describing the fluid motion is

$$\rho \frac{\partial u}{\partial t} = \mu \frac{\partial^2 u}{\partial y^2}$$

相应的水平方向（x方向）压力梯度只是x的函数，或只是y的函数，还是x和y的函数？

【答：x方向的压力梯度只是x的函数】

题 5.16 一个定常有黏不可压缩流场，忽略体积力，速度分量为

式中，a、b、c都是常数。(a) 利用纳维-斯托克斯方程求x方向压力梯度的表达式；(b) 在$y = 0$、速度为零处，求a、b、c满足什么样的关系时剪切应力$\tau_{xy} = 0$。

【答：（a）$\dfrac{\partial p}{\partial x} = 2\mu b$；
（b）$a = bc$】

题 5.17 一个弦长为2m的翼型，在海平面速度为50m/s的气流中进行测试。(a) 求雷诺数；(b) 当该翼型安装在飞机上时，以同样的速度在高度为3000m的标准大气中飞行，雷诺数会是多少？

【答：（a）6.85×10^6；
（b）5.37×10^6】

题 5.18 两个平行平板之间的不可压缩流体如图5.23所示，由于底部平板在某一时刻突然以恒定速度U移动而被带动起来，用来描述流体运动的控制微分方程为

where u is the velocity in the x-direction, and ρ and μ are the fluid density and viscosity, respectively. Rewrite the equation and the initial and boundary conditions in dimensionless form using h and U as reference parameters for length and velocity, and $h^2\rho/\mu$ as a reference parameter for time.

[Ans: $\dfrac{\partial u^*}{\partial t^*}=\dfrac{\partial u^*}{\partial y^*}$, where $u^*=u/U$, $y^*=y/h$ and $t^*=t/(h^2\rho/\mu)$]

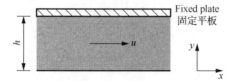

Figure 5.23　Incompressible flow between two large parallel plates

Problem 5.19　A fluid flows through a pipe of radius R with a Reynolds number of 100000. (a) At what radial location, r/R, does the fluid velocity equal the average velocity? (b) Repeat if the Reynolds number is 1000.

[Ans: (a) 0.758; (b) 0.707]

Problem 5.20　Determine the pressure drop per 300m length of 0.20m diameter horizontal cast iron pipe when the average velocity is 1.7m/s. The equivalent roughness of cast iron pipe is 0.26mm. Assume the viscosity of water to be $1.12\times10^{-3}\text{kg/(m}\cdot\text{s)}$.

[Ans: 47.685kPa]

Problem 5.21　Water at 80℃ flows through a 120mm diameter pipe with an average velocity of 2m/s. If the pipe wall roughness is small enough so that it does not protrude through the laminar sublayer, the pipe can be considered to be smooth. Approximately what is the best roughness allowed to classify the pipe as smoothness?

[Ans: 2.28×10^{-5} m]

Problem 5.22 Determine the lift and drag coefficients (based on frontal area) for the triangular two-dimensional object, shown in Figure 5.24. Assume that, skin friction is negligible.

[Ans: $C_L = 0$, $C_D = 1.7$]

认为是光滑的，那么能够使管道被划归为光滑管的最大粗糙度大约是多少？

【答：2.28×10^{-5} m】

题 5.22 计算如图 5.24 所示的二维三角形物体的升力和阻力系数（使用迎风面面积），假设可以忽略表面摩擦力。

【答：$C_L = 0$，$C_D = 1.7$】

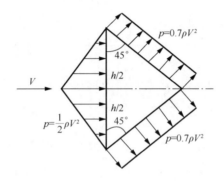

Figure 5.24 A two-dimensional triangular object

图 5.24 二维三角形物体

Problem 5.23 A horizontal pipe of length L and diameter D conveys air. Assuming the air to expand according to the law p/ρ = constant and that acceleration effects are small, prove that,

$$\frac{p_1}{\rho_1}(\rho_1^2 - \rho_2^2) = \frac{16 f L \dot{m}^2}{\pi^2 D^5}$$

where $\dot{m}$ is the mass flow rate of air through the pipe, f is the average friction coefficient, and 1 and 2 are the inlet and discharge ends of the pipe, respectively.

Problem 5.24 A viscous incompressible fluid is in a two-dimensional motion in circles about the origin with tangential velocity

$$u_\theta = \frac{1}{r} f\left(\frac{r^2}{vt}\right)$$

where v is kinematic velocity and t is time. Find

题 5.23 一个水平空气输送管的长度为 L，直径为 D。设空气膨胀遵循 p/ρ = 常数，并且加速效应可以忽略，证明

式中，$\dot{m}$ 是管中空气的质量流量；f 是平均摩擦系数；1 和 2 分别是管道的入口端和出口端。

题 5.24 一个黏性不可压缩流体，处于围绕原点的二维环形运动中，其切向速度为

式中，v 为运动黏度；t 是时

the vorticity ζ.

$$[\text{Ans:} \zeta = \frac{2}{vt}f'\left(\frac{r^2}{vt}\right)]$$

Problem 5.25 When a circulation of strength Γ is imposed on a circular cylinder placed in a uniform incompressible flow of velocity U_∞, the cylinder experiences lift. If the lift coefficient $C_L = 2$, calculate the peak (negative) pressure coefficient on the cylinder.

[Ans: -4.375]

Problem 5.26 Considering the flow field that results if a vortex with clockwise circulation is superposed on a doublet and uniform flow combination, determine the maximum lift coefficient that can be generated for a circulating flow around a cylinder, neglecting the cases of strong circulations which will shift the stagnation points outside the body.

[Ans: 4π]

Problem 5.27 A wing with an elliptical plan-form and an elliptical lift distribution has aspect ratio 6 and span of 12m. The wing loading is 900N/m^2, when flying at a speed of 150km/h at sea level. Calculate the induced drag for this wing.

[Ans: 968.3N]

Problem 5.28 A circular cylinder of radius a is in an otherwise uniform stream of inviscid fluid but with a positive circulation round the cylinder, as shown in Figure 5.25. Find the lift F and drag f per unit span of the cylinder. Also, sketch the streamlines around the cylinder if the circulation is subcritical.

[Ans: $F = \rho V_\infty \Gamma$, $f = 0$]

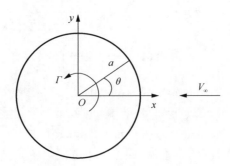

Figure 5.25　Circular cylinder in a uniform stream

图 5.25　均匀流中的圆柱

References
参考文献

[1] Hoerner S F. Fluid-Dynamic Drag: Practical Information on Aerodynamic Drag and Hydrodynamic Resistance[M]. Brick Town, New Jersey: Hoerner Fluid Dynamics, 1965.

[2] Rathakrishnan E. Theoretical Aerodynamics[M]. London: John Wiley & Sons, 2013.

Chapter 6 Boundary Layer

第 6 章 边界层

6.1 Introduction

The concept of boundary layer conceived by Ludwig Prandtl (1857-1953) in 1904 was one of the greatest inventions in the field of fluid dynamics. The concept of "fluid boundary layer" laid the foundation for the unification of the theoretical and experimental aspects of fluid mechanics. To gain an insight of this concept lets us examine the streaming flow of a fluid past a body of reasonable slender form, shown in Figure 6.1. In the majority of problems associated with aerodynamics, the fluid viscosity is relatively small, so that, unless the transverse velocity gradients are appreciable, the shearing stresses developed [given by equation $\tau = \mu(\partial u / \partial y)$] will be very small. For flows, such as that indicated in Figure 6.1, the transverse velocity gradients are usually negligibly small throughout the flow field except for thin layers of the fluid immediately adjacent to the solid body boundaries. Within these boundary layers of the fluid, however, large shearing velocities are produced with consequent shearing stresses of considerable magnitude.

Prandtl pointed out that these boundary layers were usually thin, provided the body was of streamlined form, at a moderate angle of incidence to the flow and that the flow Reynolds number was sufficiently large, so that, as a first approximation, their presence might be ignored in order to estimate the pressure field produced about the body.

6.1 引言

由路德维希·普朗特（1857—1953）于1904年提出的边界层概念是流体力学领域伟大的创造之一。这一"流体边界层"概念为流体力学中理论流体力学和实验流体力学两个分支的统一奠定了基础。为理解这一概念，让我们来观察流体流经一个细长型物体所形成的绕流流动，如图6.1所示。在大多数与空气动力学相关的问题中，流体黏度相对较小，因此，除非横向（壁面法向上）速度梯度是可观的，否则产生的剪切应力［由等式 $\tau = \mu(\partial u / \partial y)$ 给出］将会非常小。对于如图6.1所示的一些流动，除了紧挨固体边界的一薄层流场之外，流场中横向速度梯度通常都小到可以忽略。然而在这些流体边界层中，剪切速度很大，从而产生了足够大的剪切应力。

普朗特指出，假设物体是流线型的，与来流成一定攻角，并且流动雷诺数足够大，那么这些边界层通常都是很薄的。因此，为估算物体周围产生的压力场，首先应假设可以忽略边界层的存在。

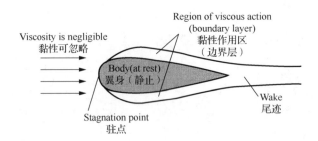

Figure 6.1 Viscous flow past an aerofoil

图 6.1 翼型黏性绕流

For aerofoil shapes, pressure field around them is only slightly modified by the boundary layer flow, since almost the entire lifting force is produced by normal pressures at the aerofoil surface. Therefore, it is possible to develop theories for the evaluation of the lift force by consideration of the flow field outside the boundary layers, where the flow is essentially inviscid in behavior. From this we can understand the importance of the inviscid flow theories. But it is important to realize that no drag force, other than induced drag, can be evaluated from inviscid flow theories. For streamlined bodies like aerofoil, the drag force is essentially due to shearing stresses at the body surface and the study of boundary layer behavior is essential for estimating these. Prandtl's boundary layer concept aids enormous simplification in the study of the whole problem. The equations of viscous motion need to be considered only in the limited regions within the boundary layers, where appreciable simplifying assumptions can reasonably be made. However, in spite of this simplification, the prediction of boundary layer behavior is still by no means simple.

由于升力几乎都是由翼型表面的正压力产生的，边界层流动对翼型周围的压力场影响很小。因此，可以通过选取边界层以外基本上无黏的流动来建立升力计算理论，由此我们会意识到无黏流动理论的重要性。但是也要意识到从无黏流动理论会推导出物体不受阻力作用。对于像翼型这样的流线型物体，阻力主要是物体表面的剪切应力引起的，因此要计算这一阻力必须要研究边界层行为。普朗特的边界层概念使整个问题的研究大大简化，只需要在边界层中能够做合理简化假设的有限区域内考虑黏性运动方程就可以。尽管有这样的简化，但对边界层流动特性的预测仍然是不容易的。

6.2 Boundary Layer Development

6.2 边界层发展

Essentially there are two types of boundary layer developments encountered in practice. They are the

实际应用中基本上有如下两种边界层发展的形式：

following.

(1) For flow around a body with a sharp leading edge, the boundary layer will grow from zero thickness at the upstream edge of the body.

(2) For a body with blunt nose, like that of a typical aerofoil, boundary layer will develop on top and bottom surfaces from the front stagnation point. The boundary layer for such bodies will have finite (not zero) thickness at the leading edge also, unlike a body with sharp leading edge for which the boundary layer thickness is zero at the leading edge.

When the flow proceeds downstream along a surface, large shearing gradients and stresses develop adjacent to the surface because of the relatively large velocities in the mainstream and the condition of no-slip (zero velocity) at the surface. Initially, this shearing action occurs only at the body surface and retards the layers of fluid adjacent to the surface, causing the fluid elements in contact with the surface to come to rest. This is popularly termed as no slip condition. These elements in turn will interact with the elements in the layers above them and retard their motion. In this way, as the fluid near the surface passes downstream, the retarding action penetrates farther away from the surface and the boundary layer of retarded fluid thickens up.

The flow velocity increases continuously from zero at the surface to nearly the freestream value at the edge of the boundary layer. Let y be the perpendicular distance from the surface at any point and let u be the corresponding velocity parallel to the surface. Consider a small element of the fluid of unit depth normal to the flow plane, having unit length in the direction of motion and a thickness δy normal to the flow direction, as shown in Figure 6.2.

（1）对于有尖锐前缘的物体的绕流，边界层厚度在物体前缘处由零开始增长。

（2）对于像典型翼型这样的钝头体，边界层从前驻点处在翼型上下表面开始发展。这类物体的边界层在前缘处也是有一定（不为零）厚度的，不像前缘尖锐物体那样边界层在前缘处厚度为零。

当流动沿物体表面流向下游时，由于主流中相对较大的流速和壁面无滑移（零速）条件形成速度差，所以紧邻物体表面会形成较大的剪切梯度和应力。最初，这一剪切运动只出现在物体表面，通过阻滞紧邻表面的流体层的流动使得与表面接触的流体微元停止流动。这就是常说的无滑移条件。之后这些流体微元反过来又会和上一层流体微元相互作用并阻滞它们的运动。这样，当物体周围的流体流向下游时，阻滞作用从物体表面向远处传递，使受阻滞流体形成的边界层增厚。

边界层中流速的分布由物体表面处的零逐渐增加到接近于边界层边缘处的来流速度。设 y 为任意一点与物体表面的垂直距离，令 u 为该点平行于物体表面的流速。取一个在垂直于流动平面方向上有单位厚度的流体微元，在运动方向上有单位长度，垂直于流动方向上有厚度 δy，如图 6.2 所示。

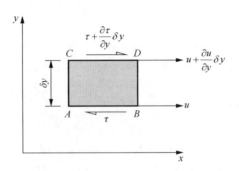

Figure 6.2　A fluid element in a viscous flow

图 6.2　黏性流动中的流体微元

The shearing stress on the face AB will be $\tau = \mu(\partial u / \partial y)$ and that on the face CD will be $\tau + (\partial \tau / \partial y)\delta y$, in the direction shown, assuming u increases with y. Thus, the resultant shearing force in the x-direction will be $[\tau + (\partial \tau / \partial y)\delta y] - \tau = (\partial \tau / \partial y)\delta y$. But $\tau = \mu(\partial u / \partial y)$, therefore, the net shear force on the element is $\mu(\partial^2 u / \partial y^2)\delta y$. Unless μ is zero, it follows that $\partial^2 u / \partial y^2$ can not be infinite and therefore, the rate of change of velocity gradient in the boundary layer must also be continuous.

Now it is clear that a smooth curve of the general form, as shown in Figure 6.3, must develop if the velocity is plotted against the distance y. Note that, at the surface ($y = 0$) the curve is not tangential to the x-axis as this would imply an infinite gradient $\partial u / \partial y$, and hence an infinite shearing stress at the surface. It is also evident that, as the shearing gradient decreases the retarding action decreases, so that at some distance from the surface, where $\partial u / \partial y$ becomes very small, the shearing stress becomes negligible, although theoretically a small gradient must exist out to $y = \infty$.

假设 u 沿 y 方向增加，AB 面上剪切应力为 $\tau = \mu(\partial u / \partial y)$，则 CD 面上的剪切应力为 $\tau + (\partial \tau / \partial y)\delta y$，作用方向如图 6.2 所示。因此，$x$ 方向上产生的剪切力应为 $[\tau + (\partial \tau / \partial y)\delta y] - \tau = (\partial \tau / \partial y)\delta y$。而 $\tau = \mu(\partial u / \partial y)$，所以微元上的总剪切力为 $\mu(\partial^2 u / \partial y^2)\delta y$。除非 μ 为零，否则 $\partial^2 u / \partial y^2$ 有极值，因此边界层中速度梯度变化率一定也是连续的。

如果相对于距离 y 画出速度，如图 6.3 所示，显然可以画出一个一般形式的光滑曲线。注意，在表面（$y = 0$）处，这一曲线和 x 轴不相切，因为那样将意味着有一个无限大的梯度值 $\partial u / \partial y$，因而在表面处将有一个无限大的剪切应力。还可以证明，阻滞运动随着剪切梯度的减小而减弱，于是在距表面某一距离处，速度梯度 $\partial u / \partial y$ 变得很小，尽管理论上在 $y = \infty$ 处也一定存在一个很小的梯度，但剪切应力已变得微乎其微。

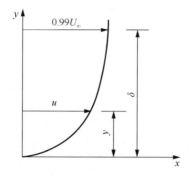

Figure 6.3　Velocity profile in a boundary layer

图 6.3　边界层中的速度分布

It will be useful to note that, the value of viscosity coefficient μ to be used in a turbulent boundary layer, will not, in general, be the simple coefficient of absolute viscosity of the fluid (see Section 2.2.4).

6.3　Boundary Layer Thickness

In order to make the concept of boundary layer realistic, an arbitrary decision must be made as to its extent and the usual convention is that the boundary layer extends to a distance δ from the surface such that the velocity u at that distance is 99 percent of the local mainstream velocity U_∞. Thus, δ is the physical thickness of the boundary layer.

The boundary layer thickness may also be defined as "the distance, from the solid boundary within which the local value of the flow velocity increases from zero at the wall ($y=0$) to 0.99 of the freestream value at the edge of the layer ($y=\delta$)", that is,

at $y=0$: $u=0$, at $y=\delta$: $u=0.99U_\infty$

where u is the local velocity, U_∞ is the freestream velocity. The above definitions of boundary layer is somewhat arbitrary.

需要注意的是，湍流边界层中使用的黏度值 μ 通常不会是简单的流体绝对黏度（见 2.2.4 节）。

6.3　边界层厚度

为使边界层概念更实用化，必须做一定的人为规定，通常的惯例是认为边界层从壁面发展到一个距壁面距离 δ 处，此处速度 u 是当地主流速度 U_∞ 的 99%。于是，δ 就是边界层的物理厚度。

边界层厚度也可定义为"距离固体边界的一段距离，在这一距离中当地流速值由壁面 $y=0$ 处的零逐渐增加，到边界层边缘 $y=\delta$ 处增加到来流速度的 0.99 倍。"即

在 $y=0$: $u=0$

在 $y=\delta$: $u=0.99U_\infty$

式中，u 是当地速度；U_∞ 是来流速度。上述边界层定义一定程度上是人为假定的。

6.3.1 Displacement Thickness

A more physically meaningful definition of the boundary layer, can be introduced by considering a parameter known as displacement thickness δ^* to account for the decrease in the total flow rate caused by the boundary layer. The displacement thickness is defined as "the distance by which the wall would have to be displaced outward in a hypothetical frictionless flow so as to maintain the same mass flux as in the actual flow".

Examine the velocity profile shown in Figure 6.4. If the boundary layer is ignored and the flow is treated as potential, the flow rate through the thickness δ would be $\rho_\infty U_\infty$ (considering the flow to be two-dimensional). But when the boundary layer is considered, the flow rate through the same area will decrease from $\rho_\infty U_\infty$ to some value ρu. Therefore, to satisfy the continuity the streamtube cross-sectional area will increase, and streamtube in the mainstream flow will be displaced slightly away from the surface. The effect on the mainstream flow will then be as if, with no boundary layer present; the solid surface has been displaced a small distance into the stream, as shown in Figure 6.4.

6.3.1 位移厚度

通过假定一个被称为位移厚度 δ^* 的参数可引出更具有物理意义的边界层定义,以计算由边界层引起的总流量下降。位移厚度解释为:为了使得假想的无摩擦流动和实际流动具有同样的质量流量,物体壁面应向外移动的距离。

观察图 6.4 所示的速度分布,如果忽略边界层并且把流动看成势流,流经厚度 δ 的流量应为 $\rho_\infty U_\infty$(假设流动是二维的)。但是考虑边界层时,经过同样面积的流量将从 $\rho_\infty U_\infty$ 减小到某一 ρu 值。因此,为满足连续性,流管横截面积应增加,或者主流中的流管应被稍微从壁面向外排挤。对主流的影响就像是没有边界层时,固体壁面向主流中移动了一段距离,如图 6.4 所示。

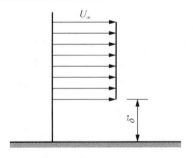

Figure 6.4 Displacement thickness

图 6.4 位移厚度

In mathematical terms, the mass of fluid which is absent due to the presence of the boundary layer is

采用数学描述,由于边界层的存在而减少的流体质量为

$\rho_\infty U_\infty \delta^*$. Equating this to the mass of the fluid which is absent due to the actual boundary layer gives the equation which defines the displacement thickness, δ^*. Thus,

$$\rho_\infty U_\infty \delta^* = \int_0^\delta (\rho_\infty U_\infty - \rho u) \mathrm{d}y$$

that is,

$$\delta^* = \int_0^\delta \left(1 - \frac{\rho u}{\rho_\infty U_\infty}\right) \mathrm{d}y \tag{6.1}$$

where ρ and u are the local density and velocity of the flow, respectively. ρ_∞ and U_∞ are the density and velocity of the freestream.

For incompressible flows, the density will be cancelled out and Equation (6.1) will be reduced to

$$\delta^* = \int_0^\delta \left(1 - \frac{u}{U_\infty}\right) \mathrm{d}y \tag{6.2}$$

The upper limit in Equation (6.2) may be allowed to extend to infinity because, $u/U_\infty \to 0$ exponentially far in y as $y \to \infty$, at the edge of the boundary layer. The concept of displacement thickness proposed here is purely based on two-dimensional flow past a flat plate in order to conceive the concept in its simplest form. The above equations may be used for any two-dimensional flow without restriction. They also can sensibly be used for three-dimensional bodies provided the curvature in the plane normal to the freestream direction is not large, that is, if the local radius of curvature is much greater than the boundary layer thickness. When the curvature is large a displacement thickness may still be defined but the form of Equations (6.1) and (6.2) will be slightly modified. The concept of the displacement thickness is used in the design of ducts, intakes of air-breathing engine, wind tunnels, etc. by first assuming a frictionless flow and then enlarging the passage wall by the displacement thickness so as to allow the same flow rate. Another use of displacement thickness is in finding the pressure gradient dp/dx at the edge of the boundary layer, needed for solving the boundary layer equations. The first approximation is to neglect the existence of the boundary layer, and calculate the irrotational dp/dx over the body surface. A solution of the boundary layer equations gives the displacement thickness, using Equation (6.2). The body surface is then displaced outward by this distance and a next approximation of dp/dx is found from a solution of the irrotational flow, and so on.

Using similar arguments to those given for boundary layer and displacement thicknesses, we can define other thicknesses associated with boundary layer, namely the momentum and energy thicknesses, using momentum and energy flow rates, respectively.

6.3.2　Momentum Thickness

A second type of boundary layer thickness which is frequently used is the momentum thickness, denoted by θ. This is defined based on the momentum flow rate within the boundary layer. This rate is less than that which would occur if no boundary layer existed, and the velocity in the vicinity of the surface at the station considered would be constant and equal to the mainstream velocity U_∞.

The momentum thickness is defined as the thickness of a layer of the fluid of velocity U_∞ for which the momentum flux is equal to the deficit of momentum flux through the boundary layer.

For a streamtube of thickness δy within the boundary layer (Figure 6.3) the rate of momentum defect (relative to the mainstream) is $\rho u(U_\infty - u)\delta y$. Note that, the actual mass flow rate ρu within the streamtube must be used here, the momentum defect of this mass being the difference between its momentum based on the mainstream velocity and its actual momentum at position x within the boundary layer.

The rate of momentum defect for the thickness θ

方程以求解所需的压力梯度 dp/dx。首先近似忽略边界层的存在并计算物体表面无旋的压力梯度 dp/dx，用式（6.2）求解边界层方程的一个解，给出位移厚度，然后把物体表面向外移动这一位移，再由无旋流动的解找到压力梯度 dp/dx 的另一个近似值，等等。

通过类似对边界层和位移厚度的讨论，我们可以用动量流量和能量流量分别定义称为动量厚度和动能厚度的关于边界层的其他厚度。

6.3.2　动量厚度

第二种常用的边界层厚度是动量损失厚度，简称动量厚度，用 θ 表示，是基于边界层内的动量流量定义的。这一流量比不存在边界层时的流量要小，所选取静止表面周围的速度是常数且等于主流速度 U_∞。

动量厚度定义为速度为 U_∞ 的一层流体的厚度，在这一厚度层中动量流量等于边界层中损失的动量流量。

对于边界层（图6.3）中厚度为 δy 的流管，损失的动量流量（相对于主流）是 $\rho u(U_\infty - u)\delta y$。注意，这里必须用流管中实际质量流量 ρu，这一质量的动量损失是主流速度的动量和边界层中某一位置 x 处实际动量的差值。

厚度 θ 的动量流量损失写

is given by $\rho_\infty U_\infty^2 \theta$. Thus,

$$\rho_\infty U_\infty^2 \theta = \int_0^\delta \rho u(U_\infty - u) \mathrm{d}y$$

or

$$\theta = \int_0^\delta \frac{\rho u}{\rho_\infty U_\infty}\left(1 - \frac{u}{U_\infty}\right) \mathrm{d}y \qquad (6.3)$$

For the case of incompressible flow this becomes

$$\theta = \int_0^\delta \frac{u}{U_\infty}\left(1 - \frac{u}{U_\infty}\right) \mathrm{d}y$$

The momentum thickness concept is conveniently used in transition problem and the calculation of skin friction losses. From the expression of θ above, we can infer that the momentum thickness is the distance through which the surface would have to be displaced in order that, with no boundary layer the total flow momentum at the station x considered would be the same as that actually occurring.

6.3.3 Kinetic Energy Thickness

The kinetic energy thickness, denoted by ϑ, may be defined as the distance through which the surface would have to be displaced in order that, with no boundary layer the total flow kinetic energy at the station considered would be the same as that actually occurring. This quantity is defined with reference to kinetic energy of the fluid in a manner comparable with the momentum thickness. The rate of kinetic energy defect within the boundary layer at any station x is given by the difference between the kinetic energy which the element would have at mainstream velocity U_∞ and that which it actually has at velocity u, being equal to

$$\int_0^\delta \frac{1}{2}\rho u(U_\infty^2 - u^2) \mathrm{d}y$$

and the rate of kinetic energy defect in the thickness ϑ is $\frac{1}{2}\rho_\infty U_\infty^3 \vartheta$. Thus,

$$\frac{1}{2}\rho_\infty U_\infty^3 \vartheta = \int_0^\delta \frac{1}{2}\rho u(U_\infty^2 - u^2)\mathrm{d}y$$

or	或

$$\vartheta = \int_0^\delta \frac{\rho u}{\rho_\infty U_\infty}\left[1-\left(\frac{u}{U_\infty}\right)^2\right]\mathrm{d}y \qquad (6.4)$$

For the case of incompressible flow, Equation (6.4) becomes	对于不可压缩流动，式（6.4）简化为

$$\vartheta = \int_0^\delta \frac{u}{U_\infty}\left[1-\left(\frac{u}{U_\infty}\right)^2\right]\mathrm{d}y$$

The displacement, momentum, and energy thicknesses can be expressed in nondimensional form, by diving them by the boundary layer thickness, to result in	位移厚度、动量厚度和动能厚度与边界层厚度相除后可用无量纲形式表示为

$$\frac{\delta^*}{\delta} = \int_0^1 \left(1-\frac{\rho u}{\rho_\infty U_\infty}\right)\mathrm{d}\overline{y}$$

$$\frac{\theta}{\delta} = \int_0^1 \frac{\rho u}{\rho_\infty U_\infty}\left(1-\frac{u}{U_\infty}\right)\mathrm{d}\overline{y}$$

$$\frac{\vartheta}{\delta} = \int_0^1 \frac{\rho u}{\rho_\infty U_\infty}\left[1-\left(\frac{u}{U_\infty}\right)^2\right]\mathrm{d}\overline{y}$$

where $\overline{y} = y/\delta$. In the above three equations the integrals on the right-hand side are simply numbers which may be evaluated readily if the boundary layer velocity profile is known.	式中，$\overline{y} = y/\delta$。在上述三个表达式中，如果边界层速度分布是已知的，右侧的积分项就是可以直接计算得到的数。

6.3.4 Non-Dimensional Velocity Profile | 6.3.4 无量纲速度分布

Expression of velocity profile in non-dimensional form will be useful to compare boundary layer profiles of different thickness. This may be done by writing $\overline{u} = u/U_\infty$ and $\overline{y} = y/\delta$, so that the velocity profile shape is given by $\overline{u} = f(\overline{y})$, where u is the local velocity and U_∞ is the freestream velocity. Over the range $y=0$ to $y=\delta$, the dimensionless velocity $\overline{u}$ varies from 0 to 0.99. For convenience when using $\overline{u}$ values as integration limits, negligible error is introduced by using	速度分布表达式采用无量纲形式将有助于比较不同厚度边界层的速度分布，只要写出 $\overline{u} = u/U_\infty$ 和 $\overline{y} = y/\delta$，则速度分布形式可表示为 $\overline{u} = f(\overline{y})$，其中 u 是当地速度，U_∞ 是来流速度。在 $y=0$ 到 $y=\delta$ 范围内，无量纲速度 $\overline{u}$ 从 0 变化到 0.99。为方便计算，把 $\overline{u}$ 值作为积分上限。当用边界层边缘 $\overline{u} = 1.0$ 而不用 0.99 来

$\bar{u} = 1.0$ instead of 0.99, at the edge of the boundary layer, and considerable arithmetic simplification is achieved. The velocity profile is then plotted as in Figure 6.5.

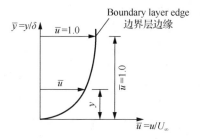

Figure 6.5 Non-dimensional velocity profile

6.3.5 Types of Boundary Layer

It has been proved experimentally that, as far as pipe flow is concerned, two different flow regimes, namely laminar and turbulent flow can exist. In laminar flow the layers of fluid elements slide smoothly over one another and there is little interchange of fluid mass between adjacent layers. The shearing friction which developed to the velocity gradient is thus entirely due to the viscosity of the fluid, that is, the momentum exchange between adjacent layers is on a molecular scale only.

In turbulent flow considerable random motion exists, in the form of velocity fluctuations both along the mean direction of flow and perpendicular to it. As a result of the latter, there are appreciable transports of mass between adjacent layers. If there is a mean velocity gradient in the flow, there will be corresponding interchanges of stream-wise momentum between adjacent layers, which will result in shearing stress between them. These shearing stresses may be of much greater magnitude than those which develop as a result of purely viscous action. The stresses due to turbulence is termed as Reynolds stresses. The velocity profile shape in a turbulent boundary layer is largely controlled by these Reynolds stresses.

Due to the difference between laminar and turbulent flow shearing stresses, the velocity profiles in the two types of boundary layers are different. Typical velocity profiles in laminar and turbulent boundary layers on a flat plate, where there is no stream-wise pressure gradient, are shown in Figure 6.6.

界层中速度分布形式主要是由这些雷诺应力决定的。

由于层流和湍流剪切应力的差别，两种类型边界层中速度分布是不同的。没有流动方向上的压力梯度时，平板上层流和湍流边界层中典型速度分布如图 6.6 所示。

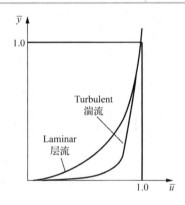

Figure 6.6　Laminar and turbulent velocity profiles in a flat plate boundary layer

图 6.6　平板边界层中的层流和湍流速度分布

In the laminar boundary layer, energy from the mainstream is transferred towards the slower moving fluid near the surface through the viscous action alone, resulting in only a relatively small perturbation. Owing to this a considerable portion of the boundary layer flow has a significantly reduced velocity. Throughout the boundary layer, the shearing stress τ is given by

在层流边界层中，主流的能量仅通过黏性运动把能量传递给接近固体壁面的运动速度更慢的流体，黏性作用只产生相对很小的扰动，因此边界层中大部分流动的流速明显降低。整个边界层中，剪切应力 τ 可写为

$$\tau = \mu \frac{\partial u}{\partial y}$$

and at the wall the shearing stress becomes

在壁面处的剪切应力变为

$$\tau_w = \mu_w \left(\frac{\partial u}{\partial y}\right)_{y=0} = \mu_w \left(\frac{\partial u}{\partial y}\right)_w$$

In the turbulent boundary layer, as already been noted, large Reynolds stresses set up due to mass transport in the direction perpendicular to the surface,

如前所述，在湍流边界层中，垂直于壁面方向的质量传递产生了很大的雷诺应力，因此主流能

so that energy from the mainstream may easily penetrate to fluid layers quite close to the surface. Because of this these layers have a velocity which is not much less than that of the mainstream. However, in layers which are very close to the surface it is not possible for the velocity to exist perpendicular to the surface, so that in a very thin region immediately adjacent to the surface, the flow approximates to laminar flow. This thin layer adjacent to the surface is termed laminar sublayer or viscous sublayer.

In the laminar sublayer the shearing action becomes purely viscous and the velocity falls very sharply, and almost linearly, within it, to zero at the surface. Therefore, the wall shear stress now depends only on viscosity, i.e. $\tau_w = \mu_w (\partial u / \partial y)_w$. The surface friction stress in a turbulent boundary layer will be far greater than that in a laminar boundary layer of the same thickness, since $(\partial u / \partial y)_w$ is much greater for turbulent boundary layer. It should be noted that, the viscous shear stress relation is employed only in the laminar sublayer very close to the surface and not throughout the turbulent boundary layer.

6.4 Boundary Layer Flow

Boundary layer may be defined as "that thin layer adjacent to a solid boundary within which the flow velocity increases from zero to 99 percent of its freestream value". Boundary layer may also be defined as that fluid layer which has had its velocity affected by the boundary shear.

Boundary layer theory is a technique to compute the viscous-layer motion near solid walls and "patch" it onto the outer inviscid flow. The patching becomes more successful as the Reynolds number becomes larger. Examine the flat plate boundary layers shown in Figures 6.7 and 6.8.

量可以很容易被传递到非常接近壁面的流体层，于是这些流体层中流速与主流流速相差不大。然而，在非常接近壁面的流体层中，由于不可能存在垂直于表面的流速，因此在紧挨壁面的一薄层区域里流动近似为层流流动。这个紧邻壁面的一薄层被称为层流底层或黏性底层。

在层流底层中，剪切作用仅是由黏性产生的，而且速度几乎线性地迅速下降至固体壁面上的零速度。因此，此时壁面剪切应力只与黏性有关，即 $\tau_w = \mu_w (\partial u / \partial y)_w$。由于湍流边界层中 $(\partial u / \partial y)_w$ 要大得多，因此湍流边界层中表面摩擦应力将远大于同样厚度层流边界层的应力。应注意，黏性剪切应力关系仅适用于接近表面的层流底层，而不是整个湍流边界层。

6.4 边界层流动

边界层可以定义为"紧邻固体边界的一个薄层，在这一层中流体流动速度从零增加到来流速度值的99%"，也可定义为速度受到了边界剪切作用影响的流体层。

边界层理论是用来计算固体壁面附近黏性层运动的一种方法，利用这一理论可以把黏性层运动和外部无黏流动"衔接"起来，雷诺数越大，这一衔接越合理。观察如图6.7和图6.8所示的平板边界层。

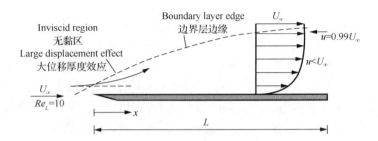

Figure 6.7 Boundary layer over a flat plate for a laminar low-Reynolds number flow

图 6.7 低雷诺数的层流平板边界层

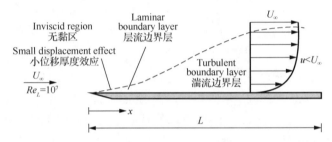

Figure 6.8 Boundary layer over a flat plate for a high-Reynolds number flow

图 6.8 高雷诺数的平板边界层

It is seen from the comparison of low- and high-Reynolds number flows over a sharp flat-plate that,

(1) For low Reynolds number, the viscous region is very broad and extends far ahead and to the sides of the plate.

(2) The plate retards the oncoming stream greatly, and small changes in flow parameters cause large changes in the pressure distribution along the plate.

(3) Thus, although in principle it should be possible to patch the viscous and inviscid layers in a mathematical analysis, their interaction is vigorous and nonlinear.

(4) There is no simple theory existing for external flow analysis for Reynolds number range from 1 to about 1000.

(5) Such thick shear layer flows are typically studied by experimental or numerical methods.

绕尖锐平板的低雷诺数和高雷诺数流动比较表明：

（1）对于低雷诺数流动，黏性区域非常厚，边界层向后一直延伸到平板的边缘。

（2）平板严重阻滞了来流速度，而且流动参数的微小变化会导致沿平板压力分布的巨大变化。

（3）因此，尽管理论上就某一问题可以把黏性层和无黏层衔接起来，但二者之间的相互作用是不规则的和非线性的。

（4）对于雷诺数在 1 到大约 1000 之间的外部流场，没有简单的理论可用于其分析。

（5）这类厚剪切层流动通常通过实验或数值计算方法来

(6) A high-Reynolds number flow is much more amenable to boundary layer patching than a low-Reynolds number flow.

(7) The viscous layers, either laminar or turbulent, are very thin, as shown in Figures 6.7 and 6.8.

(8) It can be shown for flat-plate flow that, the boundary layer thickness δ can be expressed as

$$\frac{\delta}{x} \approx \frac{5.0}{Re_x^{1/2}}$$ for laminar boundary layer

$$\frac{\delta}{x} \approx \frac{0.16}{Re_x^{1/7}}$$ for turbulent boundary layer

where $Re_x = \dfrac{\rho_\infty U_\infty x}{\mu_\infty}$ is called the local Reynolds number, ρ_∞, μ_∞, and U_∞ are the freestream density, viscosity and velocity, respectively, and x is the axial distance from the flat-plate leading edge.

The turbulent boundary layer thickness relation above applies for Reynolds number greater than approximately 10^6. Some values of δ/x with the above relations are given in Table 6.1. The blanks in Table 6.1 indicate that the relation is not applicable for those cases.

(1) In all the cases in Table 6.1, the boundary layers are so thin that their displacement effect on the outer inviscid flow region is negligible. Thus, the pressure distribution along the plate can be computed from inviscid theory as if the boundary layer is not there.

(2) For slender bodies, such as a flat-plate or an aerofoil kept parallel to the oncoming stream, the assumption of negligible interaction between the boundary layer and the outer pressure distribution is an excellent approximation.

(3) For blunt-bodies, however, even at high Reynolds number there is a discrepancy in the viscous-inviscid patching concept.

Table 6.1   Some values of δ/x

表6.1 δ/x 的一些值

Re_x	$(\delta/x)_{laminar}$	$(\delta/x)_{turbulent}$
10^4	0.050	—
10^5	0.016	—
10^6	0.005	0.022
10^7	—	0.016
10^8	—	0.011

| Examine the inviscid and actual flow fields over circular cylinders shown in Figure 6.9. | 观察图6.9所示绕圆柱的无黏和实际流场。 |

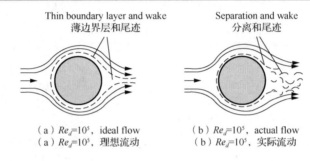

(a) $Re_d=10^5$, ideal flow
(a) $Re_d=10^5$, 理想流动

(b) $Re_d=10^5$, actual flow
(b) $Re_d=10^5$, 实际流动

Figure 6.9   Ideal flow past a circular cylinder and actual flow past a circular cylinder

图6.9   理想圆柱绕流和实际圆柱绕流

(1) In the idealized flow field there is a thin film of boundary layer about the body and a narrow sheet of viscous wake in the rear, as shown in Figure 6.9(a). The patching would be glorious for this picture, but it is false.	（1）在理想流动流场中，物体周围有一层很薄的边界层，尾部有一个狭长的黏性尾迹，如图6.9（a）所示。对于图6.9（a），这样的流场衔接是没有问题的，但却是和实际不符的。
(2) In the actual flow the boundary layer is thin on the front, or windward, side of the body, where the pressure decreases along the surface (favorable pressure gradient). But the rear boundary layer encounters increasing pressure (adverse pressure gradient) and breaks off, or separates, in a broad, pulsating wake, as shown in Figure 6.9(b).	（2）实际流动中，物体前部或迎风侧边缘的边界层很薄，沿物体表面压力逐渐降低（正压力梯度）。但尾部边界层由于压力升高（逆压力梯度）而破坏或分离，形成一个宽的脉动尾迹，如图6.9（b）所示。
(3) The theory of strong interaction between blunt-body viscous and inviscid layers is not well-developed. Flows like the actual one shown in Figure 6.9(b) are usually studied experimentally.	（3）钝体黏性层和无黏层之间相互作用的理论还不完善，因此如图6.9（b）中给出的一些实际流动，通常通过实验来研究。

6.5 Boundary Layer Solutions

For solving external flows, basically three techniques are used. They are

(1) Numerical Methods.
(2) Experimental Methods.
(3) Boundary layer theory.

Here we are concerned with only boundary layer theory. It was first formulated by Prandtl in 1904. Prandtl proposed order-of-magnitude assumptions to simplify Navier-Stokes equations to result in boundary layer equations, which can be solved relatively easily and patched onto the outer inviscid-flow filed. One of the great achievements of boundary layer theory is its ability to predict the flow separation illustrated in Figure 6.9(b). However, it should be realized that, the prediction is only approximate and not accurate.

6.6 Momentum-Integral Estimates

6.6.1 Conservation of Linear Momentum

By Newton's second law, the momentum equation for flow through a control volume can be expressed as

$$F = ma = m\frac{dV}{dt} = \frac{d}{dt}(mV)$$

where F is the force acting on the control volume, and V is the flow velocity relative to the control volume. By Reynolds transport theorem, the linear momentum relation for a deformable control-volume becomes

$$\frac{d}{dt}(mV')_{\text{system}} = \sum F = \frac{d}{dt}\left(\iiint_{cv} V'\rho d\varsigma\right) + \iint_{cs} V'\rho(V_r \cdot n)dA$$

where V' is fluid velocity relative to an inertial (non-accelerating) coordinate system, and otherwise Newton's law must be modified to include non-inertial

6.5 边界层求解

求解外部流动，基本上采用三个方法：

（1）数值方法。
（2）实验方法。
（3）边界层理论。

这里我们仅关注最初由普朗特于 1904 年建立的边界层理论。普朗特提出通过数量级假设来简化纳维-斯托克斯方程，从而得出了更易于求解同时也能够和外部无黏流场相衔接的边界层方程。边界层理论的巨大贡献之一是能够预测图 6.9（b）中的流动分离，但应该认识到这一预测只是近似的，而不是精确的。

6.6 动量积分估算

6.6.1 线性动量守恒

由牛顿第二定律，流经控制体的动量方程可表示为

式中，F 是作用于控制体的力；V 是（流体）相对于控制体的流速。由雷诺输运理论，可变控制体的线性动量关系变为

式中，V' 是相对于一个惯性（无加速）坐标系的流体流速，否则必须加入无惯性相对加速

relative acceleration terms. $\sum F$ is the vector sum of all forces acting on the control-volume material considered as a free body and V_r is the velocity of flow relative to the control-volume.

度项来修正牛顿定律；$\sum F$ 是作用在被看作是自由物体的控制体上所有力的矢量和；V_r 是相对于控制体的流速。

The entire equation is a vector relation, both the integrals are vectors due to the term V' in the integrals. The equation thus has three components. If we want only, say, the x-component, the equation reduces to

因为积分中的 V' 项是矢量，整个方程是一个矢量关系，两个积分都是矢量，因此，方程有三个分量。如果我们只想求解某一方向分量，比如 x 分量，方程简化为

$$\sum F_x = \frac{\mathrm{d}}{\mathrm{d}t}\left(\iiint_{cv} u\rho\,\mathrm{d}\varsigma\right) + \iint_{cs} u\rho(V_r \cdot n)\mathrm{d}A$$

and similarly, $\sum F_y$ and $\sum F_z$ would involve v and w, respectively.

同样地，$\sum F_y$ 和 $\sum F_z$ 将分别由 v 和 w 计算。

For a fixed control-volume, the relative velocity $V' \equiv V$, and we can use the partial derivative to express the momentum equation as

对于一个固定控制体，相对速度 $V' \equiv V$，于是我们可以用偏微分把动量方程表示为

$$\sum F = \frac{\partial}{\partial t}\left(\iiint_{cv} V\rho\,\mathrm{d}\varsigma\right) + \iint_{cs} V\rho(V \cdot n)\mathrm{d}A \qquad (6.5)$$

This is called the momentum-integral relation.

这称为动量积分关系式。

Consider a shear layer of unknown thickness grows along the sharp flat plate shown in Figure 6.10. The Drag force on the plate is given by the following momentum integral across the exit plane.

取一不计厚度沿尖锐平板增长的剪切层，如图 6.10 所示。平板受到的阻力由如下沿整个流出平面的动量积分求得：

$$D(x) = \rho b \int_0^{\delta(x)} u(U - u)\mathrm{d}y \qquad (6.6)$$

Thus, the momentum-integral equation finds application in the boundary layer analysis.

这样，就可以把动量积分方程应用到边界层分析中。

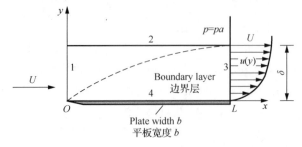

Figure 6.10 Boundary layer over a flat plate

图 6.10 平板边界层

6.6.2 Karman's Analysis of Flat Plate Boundary Layer

Equation (6.6) was derived by Karman in 1921, who wrote it in the convenient form of the momentum thickness θ, as follows.

$$D(x) = \rho b U^2 \theta$$

where the momentum thickness θ is given by

$$\theta = \int_0^{\delta(x)} \frac{u}{U}\left(1 - \frac{u}{U}\right) \mathrm{d}y \qquad (6.7\mathrm{a})$$

Thus, the momentum thickness is a measure of total drag of the plate. Karman then noted that, the drag is also equal to the integral of wall shear stress along the plate. Therefore, the drag can also be expressed as

$$D(x) = b \int_0^x \tau_w(x) \mathrm{d}x \qquad (6.7\mathrm{b})$$

or

$$\frac{\mathrm{d}D}{\mathrm{d}x} = b\tau_w$$

Differentiating the equation $D = \rho b U^2 \theta$, we get

$$\frac{\mathrm{d}D}{\mathrm{d}x} = \rho b U^2 \frac{\mathrm{d}\theta}{\mathrm{d}x}$$

Comparing the above equation with Equation (6.7), Karman arrived at what is now called the momentum-integral relation for flat-plate boundary layer, as

$$\tau_w = \rho U^2 \frac{\mathrm{d}\theta}{\mathrm{d}x} = \rho U^2 \frac{\mathrm{d}}{\mathrm{d}x} \int_0^\delta \left(1 - \frac{u}{U}\right) \mathrm{d}y \qquad (6.8)$$

This is the momentum integral equation, and is valid for both laminar and turbulent flows.

To get a numerical result for laminar flow, Karman assumed that the velocity profile had an approximately parabolic profile and expressed that as

$$u(x,y) \approx U\left(\frac{2y}{\delta} - \frac{y^2}{\delta^2}\right), \quad 0 \leqslant y \leqslant \delta(x) \qquad (6.9)$$

6.7 Boundary Layer Equations in Dimensionless Form

The boundary layer concept is based on the fact that the layer is thin. For a flat plate this means that, at any location x along the plate, the boundary layer thickness, momentum thickness and energy thickness are all smaller than x. That is, $\delta \ll x$, $\theta \ll x$ and $\vartheta \ll x$. The structure and properties of boundary layer flow depend on whether the flow is laminar or turbulent. Also, we saw that the flow velocity increases from zero to the freestream level within the boundary layer. Thus, the following two regions of flow over a body need to be considered, even if the division between them is not very sharp.

(1) A very thin layer, termed boundary layer, adjacent to the body, in which the velocity gradient normal to the surface, that is, $\partial u / \partial y$ is very large. In this region, because of this large velocity gradient, the shear stress $\tau = \mu(\partial u / \partial y)$ may assume large values, even if the viscosity μ is small.

(2) In the region outside the boundary layer, no such velocity gradient prevails. Because of the nonexistence of velocity gradient, the influence of viscosity becomes insignificant. In this region the flow is almost frictionless and can be treated as potential.

In general, the thickness of the boundary layer decreases with viscosity. That is, the boundary layer thickness decreases as the Reynolds number increases. Many exact solutions of the Navier-Stokes equations reveal that the boundary layer thickness is proportional to the square root of the kinematic viscosity, that is,

$$\delta \sim \sqrt{\nu}$$

For simplifying the Navier-Stokes equations, it is assumed that the boundary layer thickness is very small

compared to any linear dimension L of the body, that is,	体的任何直线尺寸 L，即

$$\delta \ll L$$

In this way the solutions obtained from boundary layer equations are asymptotic and apply only to very large Reynolds numbers. Now, let us simplify the Navier-Stokes equations using an estimate of the order of magnitude of each term. For a two-dimensional flow, let us begin by assuming the wall to be flat and coincides with the x-axis, and y-axis is perpendicular to the surface. Let us express the Navier-Stokes equations in dimensionless form, by referring all velocities to the freestream velocity U, and referring all linear dimensions to a characteristic length L of the body, which is so chosen that the dimensionless derivative $\partial u/\partial x$ does not exceed unity in the region under consideration. Further, the Reynolds number	但这样得到的边界层方程的解是有误差的，且只适用于大雷诺数流动。 现在，让我们通过估算每一项的数量级来简化纳维-斯托克斯方程。对于一个二维流动，我们先假设壁面是平的，与 x 轴重合，y 轴与壁面垂直。把所有速度除以来流速度 U，所有直线尺寸除以物体的特征长度 L，从而把纳维-斯托克斯方程用无量纲形式表示出来，使求解区域中的无量纲偏导 $\partial u/\partial x$ 不会大于 1。进而，假定雷诺数

$$Re = \frac{\rho UL}{\mu} = \frac{UL}{\nu}$$

is assumed to be very large. Under these assumptions, and retaining the same symbols for the dimensionless quantities as for their dimensional counterparts, we can write the Navier-Stokes equations for plane flow as follows.	很大。基于这样的假设，如果仍然用原来有量纲的变量符号表示无量纲量，则平面流动的纳维-斯托克斯方程可表示如下。
x-direction:	x 方向：

$$\frac{\partial u}{\partial t} + u\frac{\partial u}{\partial x} + v\frac{\partial u}{\partial y} = -\frac{\partial p}{\partial x} + \frac{1}{Re}\left(\frac{\partial^2 u}{\partial x^2} + \frac{\partial^2 u}{\partial y^2}\right) \quad (6.10)$$

$$\quad 1 \quad\ 1\ \ 1\quad\ \delta\ 1/\delta \qquad\quad \delta^2 \quad 1 \quad 1/\delta^2$$

y-direction:	y 方向：

$$\frac{\partial v}{\partial t} + u\frac{\partial v}{\partial x} + v\frac{\partial v}{\partial y} = -\frac{\partial p}{\partial y} + \frac{1}{Re}\left(\frac{\partial^2 v}{\partial x^2} + \frac{\partial^2 v}{\partial y^2}\right) \quad (6.11)$$

$$\quad \delta \quad\ 1\ \delta\quad \delta\ 1 \qquad\qquad \delta^2 \quad \delta \quad 1/\delta$$

Continuity:	连续性方程：

$$\frac{\partial u}{\partial x} + \frac{\partial v}{\partial y} = 0 \quad (6.12)$$

$$\quad 1 \qquad 1$$

The boundary conditions are:
$u = v = 0$, at $y = 0$ (that is, at the wall)
$u = U$, at $y \to \infty$ (that is, at the outer edge of the boundary layer)

With the assumptions made in this analysis, the dimensionless boundary layer thickness, δ / L, for which we will retain the symbol δ, is very small compared to unity, that is, $\delta \ll 1$.

Let us now estimate the order of magnitude of each term and drop the small terms to obtain the desired equations governing the boundary layer flow. The order of $\partial u / \partial x$ is 1, thus in the continuity equation $\partial v / \partial y$ is of the order 1. Hence (since $v = 0$ at the wall), in the boundary layer, v is of the order of δ. Thus, $\partial v / \partial x$ and $\partial^2 v / \partial x^2$ are also of the order of δ. Also, $\partial^2 u / \partial x^2$ is of the order 1. The orders of magnitudes are shown in Equations (6.10) to (6.12), under each term.

Let us further assume that, the unsteady acceleration $\partial u / \partial t$ is of the same order as the convective term $u \partial u / \partial x$, which means that very sudden accelerations, such as those occurring in very large pressure waves, are excluded. Also, some of the viscous terms are of the same order of magnitude as the inertia terms, at least in the proximity of the wall. Hence, some of the second-order derivatives of velocity, such as $\partial^2 u / \partial y^2$ and $\partial^2 v / \partial y^2$ must be very large near the wall. The component of velocity parallel to the wall increases from zero, at the wall, to the value 1, in the freestream, across the layer of thickness δ. Thus,

边界条件为
在 $y = 0$ 处（即壁面），$u = v = 0$
在 $y \to \infty$ 处（即边界层外边缘），$u = U$

根据这一分析中所作的假设，无量纲边界层厚度 δ / L（这里我们仍然沿用符号 δ）是远小于 1 的，即 $\delta \ll 1$。

现在我们来估算每一项的数量级，然后去掉极小项以得到理想的边界层流动方程。$\partial u / \partial x$ 的数量级是 1，因此连续性方程中 $\partial v / \partial y$ 的数量级也是 1。由于（壁面上 $v = 0$）边界层中 v 的数量级是 δ，于是，$\partial v / \partial x$ 和 $\partial^2 v / \partial x^2$ 数量级也是 δ。$\partial^2 u / \partial x^2$ 数量级也是 1。所有数量级都表示在式（6.10）～式（6.12）中的每一项下面。

进一步假设非定常加速度 $\partial u / \partial t$ 和耗散项 $u \partial u / \partial x$ 数量级相同，这就意味着不考虑例如非常大的压力波中出现的突然加速。此外，至少在壁面附近一些黏性项和惯性项也具有同样的数量级。因此，速度的一些二阶导数，例如 $\partial^2 u / \partial y^2$ 和 $\partial^2 v / \partial y^2$ 在壁面附近一定很大。平行于壁面的速度分量经过边界层厚度 δ 从壁面处的零增加到来流速度值 1。于是有

$$\frac{\partial u}{\partial y} \sim \frac{1}{\delta} \text{ and （和）} \frac{\partial^2 u}{\partial y^2} \sim \frac{1}{\delta^2}$$

$$\frac{\partial v}{\partial y} \sim \frac{\delta}{\delta} \sim 1 \text{ and （和）} \frac{\partial^2 v}{\partial y^2} \sim \frac{1}{\delta}$$

Inserting these terms in Equation (6.10) to (6.12), the 把这些项代入式（6.10）～

viscous forces in the boundary layer can become the same order of magnitude as the inertia forces, only if the Reynolds number is of the order $1/\delta^2$, that is,

$$\frac{1}{Re} = \delta^2 \tag{6.13}$$

The x-momentum equation can be simplified by neglecting $\partial^2 u/\partial x^2$ with respect to $\partial^2 u/\partial y^2$. In the continuity equation, both the terms are of equal order and remain unaltered. In the y-momentum equation, $\partial p/\partial y$ is of the order of δ. The pressure change across the boundary layer which would be obtained by integrating the y-momentum equation is of the order of δ^2, that is, it is very small. Thus, the pressure variation in the direction normal to the boundary layer is practically invariant, that is $\partial p/\partial y = 0$, and may be assumed to be equal to that at the outer edge of the boundary layer, where its value is determined by the potential flow. The pressure in the freestream outside the boundary layer is said to be "impressed" through the boundary layer, and it depends only on the coordinate x (flow direction) and on time t.

At the outer edge of the boundary layer the parallel component of velocity u becomes equal to the velocity in the outer flow, $U(x,t)$. Since there is no large velocity gradient here, the viscous terms in Equation (6.10) vanish for large values of Re, and consequently, for the outer flow, we have

$$\frac{\partial U}{\partial t} + U\frac{\partial U}{\partial x} = -\frac{1}{\rho}\frac{\partial p}{\partial x} \tag{6.14}$$

Here again the symbols denote the dimensionless quantities.

For a steady flow the equations can be simplified further if the pressure depends only on x, that is, if the derivative $\partial p/\partial x$ becomes $\mathrm{d}p/\mathrm{d}x$, so that

变为 dp/dx，则

$$p + \frac{1}{2}\rho U^2 = \text{constant}（常数）\tag{6.15}$$

The boundary conditions for the external flow are nearly the same as for frictionless flow. The boundary layer thickness is very small and the transverse velocity component v is also very small at the edge of the boundary layer ($v/U \sim \delta/L$). Thus the potential flow about the body under consideration, in which the perpendicular velocity component is vanishingly small near the wall, offers a very good approximation to the actual external flow. The pressure gradient along the x-direction in the boundary layer can be obtained simply by analyzing the Bernoulli equation (Equation (6.15)) along the streamline at the well-known potential flow.

With the above implications and assumptions, we can write down the simplified Navier-Stokes equations, known as the Prandtl's boundary layer equations. In dimensional form, for a steady, two-dimensional, incompressible, viscous flow over a flat plate, shown in Figure 6.11, we have the boundary layer equations as

外部流动边界条件几乎和无摩擦流动是一样的，边界层厚度很小，在边界层边缘 ($v/U \sim \delta/L$) 处横向速度分量 v 也很小。这样所求解的物体周围法向速度分量在接近壁面时几乎为零的势流假设，为实际外部流动提供了很好的近似。边界层中 x 方向的压力梯度可仅通过分析已知势流处沿流线的伯努利方程［式（6.15）］求得。

由上述推理和假设，我们可以写出简化的纳维-斯托克斯方程，即普朗特的边界层方程。对于一个绕平板的定常、二维、不可压缩、黏性流动，如图 6.11 所示，有量纲形式的边界层方程为

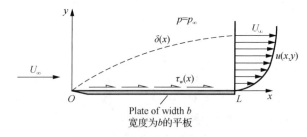

Figure 6.11　Boundary layer on a flat plate

图 6.11　平板边界层

$$\rho\left(u\frac{\partial u}{\partial x} + v\frac{\partial u}{\partial y}\right) = -\frac{\partial p}{\partial x} + \mu\left(\frac{\partial^2 u}{\partial x^2} + \frac{\partial^2 u}{\partial y^2}\right)\tag{6.16}$$

$$\frac{\partial u}{\partial x} + \frac{\partial v}{\partial y} = 0 \tag{6.17}$$

with boundary conditions

at $y = 0$(在 $y = 0$ 处)，$u = 0, v = 0$

at $y \to \infty$（在 $y \to \infty$ 处），$u \to U_\infty$

边界条件为

We know that the boundary layer is very thin. Therefore, we can assume that	由于边界层很薄，因此可以假设

$$v \ll u \quad (6.18a)$$

and	并且

$$\frac{\partial}{\partial x} \ll \frac{\partial}{\partial y} \quad (6.18b)$$

Applying these approximations to the y-component of Navier-Stokes equation [Equation (6.11)], we get	把这一近似用于纳维-斯托克斯方程的 y 方向分量，即式（6.11），有

$$\frac{\partial p}{\partial y} \approx 0 \quad (6.19a)$$

or	或

$$p \approx p(x) \quad (6.19b)$$

In other words, the y-momentum equation can be neglected completely, and the pressure p varies along the boundary layer and not through it. Equation (6.19a) implies that pressure is approximately uniform across the boundary layer. The pressure at the body surface is therefore equal to that at the edge of the boundary layer. Therefore, the pressure can be found from the solutions of the irrotational flow around the body. That is, the pressure in the outer flow at the edge of the boundary layer is "impressed" through the boundary layer. This justifies the experimental fact that the observed surface pressure on a model is approximately equal to that calculated with the irrotational (potential) flow theory. Here it is important to note that, the assumption that $\partial p / \partial y$ is vanishingly small is not valid if the boundary layer separated or about to separate from the wall or if the radius of curvature of the surface is not large compared to the boundary layer thickness.	换句话说，y 方向动量方程可以完全忽略，压力 p 只沿边界层长度方向变化，而厚度方向不变。式（6.19a）表明边界层厚度方向上的压力几乎是一致的，因此物体表面压力等于边界层边缘压力，于是求解物体周围的无旋流动就可得到其压力，即边界层边缘处外部流动中的压力是"贯穿"边界层厚度方向的。这也验证了一个实验事实，即观测到的模型壁面上压力几乎等于通过无旋（有势）流动理论计算的压力。这里尤其要指出，如果边界层有分离或者将要从壁面分离或表面的曲率半径与边界层厚度相比不够大，则 $\partial p / \partial y$ 小到可以忽略的假设不成立。
The pressure gradient $\partial p / \partial x$ in Equation (6.16) is assumed to be known in advance from Bernoulli's equations applied to the outer inviscid flow and it is	假定式（6.16）中压力梯度 $\partial p / \partial x$ 是由外部无黏流动的伯努利方程事先求得的，为

$$\frac{\partial p}{\partial x} = \frac{\mathrm{d} p}{\mathrm{d} x} = -\rho U \frac{\mathrm{d} U}{\mathrm{d} x} \quad (6.20)$$

where $U = U_\infty$, the subscript is dropped for simplicity. Further, in Equation (6.16),

$$\frac{\partial^2 u}{\partial x^2} \ll \frac{\partial^2 u}{\partial y^2}$$

Note: ① neither term in the continuity Equation (6.17) can be neglected, and ② continuity equation is always a vital part of any fluid flow analysis.

The governing Equations (6.16) and (6.17) reduce to Prandtl's two boundary layer equations given below.

Continuity equation

$$\frac{\partial u}{\partial x} + \frac{\partial v}{\partial y} = 0 \tag{6.21a}$$

Momentum equation along the wall

$$u\frac{\partial u}{\partial x} + v\frac{\partial u}{\partial y} \approx U\frac{dU}{dx} + \frac{1}{\rho}\frac{\partial \tau}{\partial y} \tag{6.21b}$$

where

$$\tau = \mu\frac{\partial u}{\partial x} \quad \text{(for laminar flow)}$$

and

$$\tau = \mu\frac{\partial u}{\partial x} - \rho\overline{u'v'} \quad \text{(for turbulent flow)}$$

The additional stress $\rho\overline{u'v'}$ in this stress expression is known as the apparent or virtual stress of turbulent flow or Reynolds stress. This is due to turbulent fluctuations which are given by the time-mean values of the quadratic terms in the turbulent components. Since this stress is added to the ordinary viscous term in laminar flow, it has similar influence on the course of the flow; it is often said that it is caused by eddy viscosity. In general, the apparent stresses far outweigh the viscous components and, consequently, the latter may be omitted in many actual cases with a good degree of approximation.

These equations are to be solved for $u(x, y)$ and

$v(x, y)$, with the assumption that $U(x)$ is known from the inviscid-flow analysis exterior of the boundary layer.

There are two boundary conditions on u and one on v. They are:

(1) At $y = 0$ (at the wall): $u = v = 0$, due to no-slip condition.

(2) At $y = \delta(x)$ exterior flow (outer stream): $u = U(x)$ by patching.

Unlike the Navier-Stokes equations which are mathematically elliptic, the boundary layer equations are parabolic and are solved by beginning at the leading edge and marching downstream as far as one likes, stopping at the separation point or earlier if it is preferred.

The boundary layer equations have been solved for many interesting cases of internal and external flows, for both laminar and turbulent conditions, utilizing the inviscid distribution $U(x)$ appropriate to each flow.

Note: the estimation of the boundary layer thickness in Equation (6.13) shows that

$$\frac{\delta}{L} \sim \frac{1}{\sqrt{Re_L}} \sim \sqrt{\frac{v}{UL}} \qquad (6.22)$$

The fact that $\delta \sim \sqrt{v}$, inferred from the exact solution of the Navier-Stokes equations, is thereby confirmed. The numerical coefficient still missing in Equation (6.22), will turn out to be equal to 5 for the case of flat plate at zero incidence, when L will mean the distance from the leading edge. Let us have a closer look at one such solution namely, the flat-plate solution in the following section.

6.8 Flat Plate Boundary Layer

One of the classic and most often used solutions of

boundary layer theory is the semi-infinite flat plate solution.

6.8.1 Laminar Flow Boundary Layer

For laminar flow past a flat-plate, the boundary layer Equations (6.16) - (6.18) can be solved exactly for u and v, assuming that the freestream velocity U is constant $(\mathrm{d}U/\mathrm{d}x = 0)$. The solution was given by Prandtl's student Blasius in his doctoral dissertation from Gottingen in 1908. With an ingenious coordinate transformation, Blasius showed that the dimensionless velocity profile u/U is a function of the single composite dimensionless variable $y\left(\dfrac{U}{\nu x}\right)^{1/2}$ only and expressed the profile as

$$\frac{u}{U} = f'(\eta), \quad \eta = y\left(\frac{U}{\nu x}\right)^{1/2} \quad (6.23)$$

where the prime denotes differentiation with respect to η. It is important here to understand how the dimensionless group for η is arrived at. Basically, the velocity ratio is a function of v, x and t and can be expressed as

$$\frac{u}{U} = f(v, x, t)$$

where the dimensions of the kinematic viscosity coefficient ν is L^2/T, the only nondimensional group which can be formed from ν, x, t is $x/\sqrt{\nu t}$ or some power thereof. The quantity η, which is a combination of the old independent variables x and t is called a similarity variable. It now plays the role of a single independent variable for the problem. Substitution of Equation (6.23) into the boundary layer Equation (6.16) reduces the problem, after much algebra, to a single third-order nonlinear ordinary equation of f, as

$$f''' + \frac{1}{2} f f'' = 0 \quad (6.24\mathrm{a})$$

The boundary conditions become	边界条件为

$$\text{at } y = 0 \text{ (在 } y=0 \text{ 处)}, \quad f(0) = f'(0) = 0 \qquad (6.24b)$$

$$\text{at } y \to \infty \text{ (在 } y \to \infty \text{ 处)}, \quad f'(\infty) \to 1.0 \qquad (6.24c)$$

In this example, both partial differential Equations (6.16) and (6.17) have been transformed into an ordinary differential equation by the substitution of $\eta = (U/\nu x)^{1/2}$. The resulting differential equation is nonlinear and of the third order. The three boundary conditions in Equation (6.24a) are, therefore, sufficient to determine the solution completely. This is the Blasius equation for which accurate solutions have been obtained only by numerical integration. Some tabulated values of the velocity-profile shape $f'(\eta) = u/U$ are given in Table 6.2.	在这一例子中，通过代入 $\eta = (U/\nu x)^{1/2}$，两个偏微分方程（6.16）与式（6.17）转化为一个常微分方程，得到的微分方程是三阶非线性的。这样式（6.24a）中的三个边界条件就足以确定方程的解。这就是只通过数值积分来求解精确解的布拉休斯方程。表 6.2 中列出了一些速度分布型 $f'(\eta) = u/U$ 的制表值。
Because u/U approaches 1.0 only as $y \to \infty$, it is customary to select the boundary layer thickness δ as that point where $u/U = 0.99$. From the table, it is seen that this occurs at $\eta \approx 5.0$. Therefore, we can write that	因为只有当 $y \to \infty$ 时，u/U 的值才接近 1.0，所以通常选择 $u/U = 0.99$ 点处为边界层厚度 δ。从表 6.2 可以看出，这出现在 $\eta \approx 5.0$ 处。因此，我们可以写出

$$\delta \left(\frac{U}{\nu x} \right)^{1/2} \approx 5.0$$

Table 6.2　The Blasius velocity profile
表 6.2　布拉休斯速度分布

$y(U/\nu x)^{1/2}$	u/U	$y(U/\nu x)^{1/2}$	u/U
0.0	0.0	2.8	0.81152
0.2	0.06641	3.0	0.84605
0.4	0.13277	3.2	0.87609
0.6	0.19894	3.4	0.90177
0.8	0.26471	3.6	0.92333
1.0	0.32979	3.8	0.94112
1.2	0.39378	4.0	0.95552
1.4	0.45627	4.2	0.96696
1.6	0.51676	4.4	0.97587
1.8	0.57477	4.6	0.98269
2.0	0.62977	4.8	0.98779
2.2	0.68132	5.0	0.99155
2.4	0.72899	⋮	⋮
2.6	0.77246	∞	1.00000

or	或者
$$\frac{\delta}{x} = \frac{5.0}{Re_x^{1/2}} \qquad (6.25)$$	
With the profile known, Blasius of course could also compute the wall shear stress coefficient, C_f as	通过已知的速度分布，布拉休斯当然还可以计算出壁面剪切应力系数C_f，为
$$C_f = \frac{0.664}{Re_x^{1/2}} \qquad (6.26a)$$	
This is the local skin friction coefficient for flow on one side of the plate. If the flow remains laminar over a length L of the plate, the total skin friction coefficient (for flow over both the surfaces of the plate) becomes	这就是平板单侧流动的当地表面摩擦系数。如果在长L的平板上始终保持层流流动，则总表面摩擦系数（对于流过平板两侧的流动）变为
$$C_f = \frac{1.328}{Re_x^{1/2}} \qquad (6.26b)$$	
Here $Re_x = \rho Ux/\mu$ denotes the Reynolds number with respect to the length of the plate and the freestream velocity. This law of friction deduced by Blasius is valid only in the region of laminar flow, that is for $Re_x < 5 \times 10^5$. In the region of turbulent flow with $Re_x > 10^6$, the drag becomes considerably greater than that given by Equation (6.26b).	此处$Re_x = \rho Ux/\mu$表示与板长和来流速度相关的雷诺数。由布拉休斯推导的这一摩擦定律只适用于层流区域，即当$Re_x < 5 \times 10^5$时成立。在湍流区域$Re_x > 10^6$时，阻力值远大于式（6.26b）的计算结果。

6.8.2　Boundary Layer Thickness for Flat Plate

6.8.2　边界层厚度

It is impossible to indicate the boundary layer thickness precisely, since the influence of velocity on the boundary layer decreases asymptotically. The parallel component of velocity u tends asymptotically to the freestream value U of the potential flow. If it is desired to define the boundary layer thickness as the distance for which $u = 0.99U$, then as seen from Table 6.2, $\eta \approx 5.0$. Hence the boundary layer thickness, as defined here, becomes

由于流动速度对边界层的影响是逐渐减小的，因此不可能准确标示出边界层厚度。速度u的水平分量渐近地趋近于势流的来流速度值，如果希望定义边界层厚度为$u = 0.99U$处距壁面距离，那么由表6.2可见，$\eta \approx 5.0$。于是像此处定义的，边界层厚度变为

$$\delta \approx 5\sqrt{\frac{vx}{U}}$$

A meaningful physical measure of the boundary layer

一个有物理意义的边界层厚度

thickness is the displacement thickness δ^*, which is earlier introduced in Equation (6.1). The displacement thickness is the distance by which the external potential field of the flow is displaced outwards as a consequence of the decrease in velocity in the boundary layer. By Equation (6.1), we have	的度量是在之前式（6.1）中给出的位移厚度 δ^*。位移厚度是由边界层中速度减小而使外部势流流场被移向外的距离，由式（6.1）有

$$\delta^* = \int_{y=0}^{\infty}\left(1 - \frac{\rho u}{\rho_\infty U_\infty}\right)\mathrm{d}y$$

With u/U from Equation (6.23), we can express this as	由式（6.23）中 u/U，上式可表示为

$$\delta^* = \sqrt{\frac{vx}{U}}\int_{y=0}^{\infty}(1 - f'(\eta))\mathrm{d}\eta$$

This can be solved to obtain	求解得到

$$\delta^* = 1.721\frac{vx}{U}$$

or	或

$$\frac{\delta^*}{x} = \frac{1.721}{(Re_x)^{1/2}} \quad (6.27)$$

When C_f is converted into dimensional form, we have the shear stress as	当 C_f 转化为量纲形式时，得到剪切力为

$$\tau_w(x) = \frac{1}{2}\rho U^2 C_f \quad (6.28a)$$

Substituting for C_f, from Equation (6.26a), we can express the shear as	在式（6.26a）中代入 C_f，写出剪切应力表达式为

$$\tau_w(x) = \frac{0.332\rho^{1/2}\mu^{1/2}U^{1.5}}{x^{1/2}} \quad (6.28b)$$

The shear stress decreases with $x^{1/2}$, a result of the thickening of the boundary layer and the associated decrease of the velocity gradient because of boundary layer growth and varies as velocity to the 1.5 power. This is in contrast to laminar pipe flow, where $\tau_w \propto U$ and is independent of x. Note that, the wall shear stress at the leading edge is predicted to be infinite[Equation(6.28b)]. Clearly the boundary layer theory breaks down near the leading edge where the assumption $\partial/\partial x \ll \partial/\partial y$ is not valid. The local Reynolds number Re_x in the neighborhood of the	可见剪切应力随 $x^{1/2}$、边界层的增厚以及由于边界层增长而导致的速度梯度相应减小而减小，并且与速度成 1.5 次方关系，这与层流管道流动 $\tau_w \propto U$ 且 τ_w 与 x 无关的情况恰恰相反。注意：前缘壁面剪切应力预计为无限大[式（6.28b）]。显然，在 $\partial/\partial x \ll \partial/\partial y$ 假设不适用的前缘附近，边界层理论也不成立。前缘附近当地雷诺数 Re_x 数量

leading edge is of the order 1, for which the boundary layer assumptions are not valid.

If τ_w is substituted into Equation (6.7a), the total drag force on one side of a plate of length x and width b becomes

$$D(x) = b\int_0^x \tau_w(x)\mathrm{d}x$$

Substituting Equation (6.28b) for τ_w, we get the drag as

$$\begin{aligned}D(x) &= \int_0^x 0.332b\rho^{1/2}\mu^{1/2}U^{1.5}x^{-1/2}\mathrm{d}x \\ &= 0.332b\rho^{1/2}\mu^{1/2}U^{1.5}\int_0^x x^{-1/2}\mathrm{d}x \\ &= 0.664b\rho^{1/2}\mu^{1/2}U^{1.5}x^{1/2}\end{aligned} \quad (6.29)$$

The drag increases only as the square root of the plate length. The drag coefficient is defined as

$$C_D = \frac{2D(L)}{\rho U^2 bL}$$

Substituting Equation (6.29), we get

$$C_D = \frac{1.328}{Re_L^{1/2}}$$

From Equation (6.26a), we get

$$C_D = 2C_f(L) \quad (6.30)$$

Thus, for laminar flat plate flow, the drag coefficient is equal to twice the value of the skin-friction coefficient at the trailing edge. This is the drag on one side of the plate. Further, Karman pointed out that the drag could be computed from the momentum relation given by Equation (6.6). In dimensionless form, Equation (6.6) becomes

$$C_D = \frac{2}{L}\int_0^\delta \frac{u}{U}\left(1-\frac{u}{U}\right)\mathrm{d}y \quad (6.31)$$

This can be written in terms of θ at the trailing edge as

$$C_D = \frac{2\theta(L)}{L} \quad (6.32)$$

Computation of θ from the profile u/U or from C_D gives

$$\theta_x = \frac{0.664}{Re_x^{1/2}} \quad \text{(laminar flat plate,层流平板边界层)} \tag{6.33}$$

Since the boundary layer thickness δ is not properly defined, the momentum thickness, being definite, is often used to correlate data taken for a variety of boundary layers under differing conditions. The ratio of displacement thickness, (δ^*), to momentum thickness, (θ), called the dimensionless- profile shape factor, H, is also useful in integral theories.

For laminar flat plate boundary layer,

由于边界层厚度 δ 没有被准确地定义,动量厚度(作为准确的定义)常用以修正许多不同条件下边界层的数据。位移厚度(δ^*)与动量厚度(θ)的比值称为无量纲速度分布形状因子 H,在积分理论中也是有用的。

对于层流平板边界层,

$$H = \frac{\delta^*}{\theta} = \frac{1.721}{0.664} = 2.59 \tag{6.34}$$

As we shall see, a large shape factor implies that the boundary layer separation is about to occur.

The plot of Blasius velocity profile from Table 6.2, in the form of u/U versus y/δ, shown in Figure 6.12, will highlight the importance of the simple guess of the velocity profile in integral theory.

我们将会看到,形状因子大则预示着边界层将要分离。

由表 6.2,以 u/U-y/δ 形式画出布拉休斯速度分布,如图 6.12 所示。这一分布图会突显出大致估算速度分布这一方法在积分理论中的重要性。

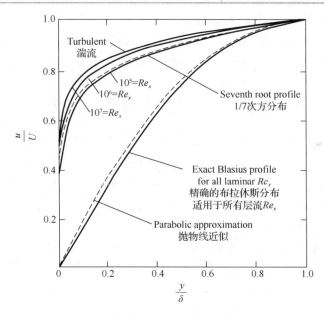

Figure 6.12 Comparison of laminar and turbulent flat plate velocity profiles

图 6.12 层流与湍流平板速度分布比较

It is seen from this plot that, the simple parabolic profile is not far from the true Blasius profile, hence its momentum thickness is within 10 per cent of the true value. Instead of decreasing monotonically to zero, the turbulent profiles are flat and then drop off sharply at the wall. As you might have guessed, they follow the logarithmic-law shape and thus can be analyzed by momentum-integral theory, if this shape is properly represented.

The flow in the boundary layer along a wall becomes turbulent when the external velocity is sufficiently large. Experimental investigation into the transition from laminar to turbulent flow in the boundary layer revealed that the transition becomes most clearly discernible by sudden and large increase in the boundary layer thickness and in the shearing stress near the wall, when x is replaced by the local length l, the dimensionless boundary layer thickness $\delta/\sqrt{(\nu x)/U}$ becomes constant in laminar flow, and is approximately equal to 5. Boundary layer thickness variation with Reynolds number, based on the current length x, along a plate in parallel flow at zero incidence is shown in Figure 6.13.

At $Re_x > 3.2 \times 10^5$, a sharp increase in the boundary layer thickness is clearly seen and an identical increase would also be experienced by the wall shear stress. The sudden increase in these quantities denotes that the flow has changed from laminar to turbulent. The Re_x, based on the local length x, is related to the Reynolds number $Re_\delta = U\delta/\nu$, based on the boundary layer thickness, through the equation

$$Re_\delta = 5\sqrt{Re_x}$$

Hence the critical Reynolds number for the plate based on the local length x,

从图 6.12 中可以看出，简单抛物线分布与真实布拉休斯分布相差不大，因此它的动量厚度应在它真实值的 10%变化范围以内。平板边界层湍流速度分布不是单调地降为0的，而是先平缓地减小，然后到壁面处突然降低。正如猜想的那样，湍流分布符合对数规律形状，因此如果正确给出了分布形状，可以用动量积分理论进行分析。

当外部流速足够大时，沿壁面的边界层流动变为湍流流动。当用局部板长 l 取代水平距离 x，无量纲边界层厚度 $\delta/\sqrt{(\nu x)/U}$ 变为层流流动中一个近似为 5 的常数时，边界层中层流到湍流转捩的实验研究揭示了转捩明显是由边界层厚度和近壁面的剪切应力突然显著增加所导致的。边界层厚度随雷诺数的变化如图 6.13 所示，该雷诺数由水平流动中零倾角平板上的当前长度 x 计算得到。

当 $Re_x > 3.2 \times 10^5$ 时，可以明显看到边界层厚度的急剧增加，同时壁面剪应力也会明显增大。这些量的突然增大标志着流动从层流转变为了湍流。这一基于当地板长 x 的 Re_x 和基于边界层厚度的雷诺数 $Re_\delta = U\delta/\nu$ 有如下关系：

因此当取边界层厚度 δ 为雷诺数表达式中长度时，基于当地板长 x 的临界雷诺数

$$Re_{x,\text{cri}} = \left(\frac{Ux}{\nu}\right)_{\text{cri}}$$

becomes approximately 2800 when the boundary layer thickness δ is taken as the length in the Reynolds number expression. The numerical value of $Re_{x,\text{cri}}$ depends on the amount of disturbance in the external flow, and the value $Re_{x,\text{cri}} = 3.2 \times 10^5$ should be regarded as the lower limit. For exceptionally disturbance-free external flow, $Re_{x,\text{cri}}$ as high as 10^6 has been obtained.

大约变为 2800。$Re_{x,\text{cri}}$ 的数值取决于外部流动的扰动程度，其最小值应取为 $Re_{x,\text{cri}} = 3.2 \times 10^5$。对于特殊的无扰动外部流动，$Re_{x,\text{cri}}$ 可以高达 10^6。

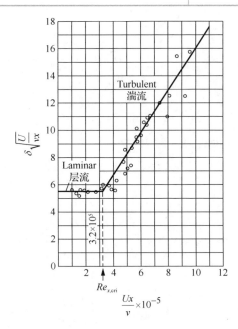

Figure 6.13　Boundary layer thickness variation in laminar and turbulent flow regimes[1]

图 6.13　边界层厚度在层流和湍流区域的变化[1]

Example 6.1

A flat plate of length 0.8m and width 1.9m is kept in sea level air stream at a velocity of 5.3m/s. Assuming a linear velocity profile for the boundary layer over the plate, develop an expression for the variation of wall shear stress with distance along the plate. Also, obtain an expression for the total skin-friction drag on the plate and evaluate the skin-friction drag. Assume that the density of air is 1.225kg/m^3, kinematic viscosity is

例 6.1

长 0.8m、宽 1.9m 的平板置于海平面风速为 5.3m/s 的气流中。假设平板表面边界层速度分布为线性分布，推导出壁面剪切应力随平板上位置变化的表达式，以及平板上总表面摩擦力的表达式，并估算表面摩擦力的值。假设空气密度为

$1.46\times10^{-5}\,\mathrm{m^2/s}$.

Solution

The velocity profile is linear, thus,

$$\frac{u}{U} = \frac{y}{\delta} = \eta$$

Hence, the momentum thickness becomes,

$$\theta = \delta \int_0^1 \frac{u}{U}\left(1 - \frac{u}{U}\right)\mathrm{d}\left(\frac{y}{\delta}\right)$$

$$= \delta \int_0^1 \eta(1-\eta)\mathrm{d}\eta$$

$$= \delta \left[\frac{\eta^2}{2} - \frac{\eta^3}{3}\right]_0^1$$

$$= \frac{\delta}{6}$$

For $u/U = y/\delta$, the wall shear stress becomes

$$\tau_w = \mu \left.\frac{\mathrm{d}u}{\mathrm{d}y}\right|_{y=0} = \mu \frac{U}{\delta}$$

The shear stress is also given, by Equation (6.8), as

$$\tau_w = \rho U^2 \frac{\mathrm{d}\theta}{\mathrm{d}x}$$

But $\theta = \delta/6$, therefore,

$$\tau_w = \mu \frac{U}{\delta} = \rho U^2 \frac{1}{6}\frac{\mathrm{d}\delta}{\mathrm{d}x}$$

Thus, we have

$$\mu \frac{U}{\delta} = \rho U^2 \frac{1}{6}\frac{\mathrm{d}\delta}{\mathrm{d}x}$$

Separating the variables and integrating, we get

$$\delta\mathrm{d}\delta = 6\frac{\mu}{\rho U}\mathrm{d}x$$

$$\frac{\delta^2}{2} = 6\frac{\mu}{\rho U}x + c$$

But $\delta = 0$ at $x = 0$, thus, $c = 0$. Therefore,

$$\delta = \sqrt{\frac{12\mu x}{\rho U}} = 3.46\left(\frac{\mu x}{\rho U}\right)^{1/2}$$

$$\frac{\delta}{x} = \frac{3.46}{\sqrt{Re_x}}$$

$$\tau_w = \mu \frac{U}{\delta}$$

$$= \frac{1}{3.46} \frac{\mu^{1/2} \rho^{1/2} U^{3/2}}{x^{1/2}}$$

$$= 0.289 \frac{U\mu}{x} \sqrt{Re_x}$$

Thus, the skin friction drag becomes	于是表面摩擦阻力变为

$$D_f = \int \tau_w \mathrm{d}A_s$$
$$= \int_0^L \tau_w (b\mathrm{d}x)$$
$$= b\int_0^L \rho U^2 \frac{\mathrm{d}\theta}{\mathrm{d}x}\mathrm{d}x$$
$$= b\int_0^L \rho U^2 \mathrm{d}\theta$$
$$= \rho U^2 b\theta_L$$

where $\theta_L = \delta_L/6$, $\delta_L = 3.46L/\sqrt{Re_L}$, b and L are the width and length of the plate, respectively. The Reynolds number is	式中，$\theta_L = \delta_L/6$，$\delta_L = 3.46L/\sqrt{Re_L}$；$b$ 和 L 分别为平板的宽与长。雷诺数为

$$Re_L = \frac{UL}{\nu}$$
$$= \frac{5.3 \times 0.8}{1.46 \times 10^{-5}}$$
$$= 2.9 \times 10^5$$

Thus,	因此

$$\delta_L = \frac{3.46 \times 0.8}{\sqrt{2.9 \times 10^5}} = 5.14\mathrm{mm}$$

$$\theta_L = \frac{\delta_L}{6} = 0.857\mathrm{mm}$$

$$D_f = \rho U^2 b\theta_L$$
$$= 1.225 \times 5.3^2 \times 1.9 \times (0.857 \times 10^{-3})$$
$$= 0.056\mathrm{N}$$

Example 6.2 The velocity profile in a laminar boundary layer flow at zero pressure gradient is approximated as	例 6.2 在压力梯度为零时，层流边界层流动中速度分布近似为

$$\frac{u}{U} = \frac{3}{2}\left(\frac{y}{\delta}\right) - \frac{1}{2}\left(\frac{y}{\delta}\right)^3$$

Obtain expressions for $\frac{\delta}{x}$ and C_f, using the momentum integral equation.	利用动量积分方程推导出 $\frac{\delta}{x}$ 与 C_f 的表达式。
Solution	**解**
From Equation (6.8), the shear stress can be expressed as	由式（6.8），剪切应力可表示为

$$\tau_w = \rho U^2 \frac{\mathrm{d}\theta}{\mathrm{d}x}$$

$$= \rho U^2 \frac{\mathrm{d}}{\mathrm{d}x}\int_0^\delta \frac{u}{U}\left(1 - \frac{u}{U}\right)\mathrm{d}y$$

$$= \rho U^2 \frac{\mathrm{d}\delta}{\mathrm{d}x}\int_0^1 \frac{u}{U}\left(1 - \frac{u}{U}\right)\mathrm{d}\left(\frac{y}{\delta}\right)$$

$$= \rho U^2 \frac{\mathrm{d}\delta}{\mathrm{d}x}\int_0^1 \left(\frac{3}{2}\eta - \frac{1}{2}\eta^3\right)\left(1 - \frac{3}{2}\eta + \frac{1}{2}\eta^3\right)\mathrm{d}\eta$$

where $\frac{y}{\delta} = \eta$. On integration, this yields	式中，$\frac{y}{\delta} = \eta$。积分得到

$$\tau_w = \frac{39}{280}\rho U^2 \frac{\mathrm{d}\delta}{\mathrm{d}x}$$

The shear stress can also be expressed as	剪切应力还可表示为

$$\tau_w = \mu \frac{\partial u}{\partial y}\bigg|_{y=0}$$

$$= \mu \frac{U}{\delta}\frac{\partial(u/U)}{\partial(y/\delta)}$$

$$= \mu \frac{U}{\delta}\frac{\partial\left(\frac{3}{2}\eta - \frac{\eta^3}{2}\right)}{\partial \eta}$$

$$= \frac{3}{2}\mu \frac{U}{\delta} \quad \text{(neglect the item of } \eta^2\text{，忽略 } \eta^2 \text{项)}$$

Thus, we have	于是有

$$\frac{3}{2}\mu \frac{U}{\delta} = \frac{39}{280}\rho U^2 \frac{\mathrm{d}\delta}{\mathrm{d}x}$$

$$\delta \mathrm{d}\delta = \frac{280}{39}\times\frac{3}{2}\frac{\mu}{\rho U}\mathrm{d}x$$

$$\frac{\delta^2}{2} = 10.77\frac{\mu}{\rho U}x + c$$

But at $x=0$, $\delta=0$, therefore, $c=0$. Thus,	而在 $x=0$ 处 $\delta=0$，因此得到

	$c=0$。于是
	$$\frac{\delta^2}{2} = \frac{21.54}{Re_x}$$ $$\frac{\delta}{x} = \frac{4.64}{\sqrt{Re_x}}$$
The skin friction coefficient is	于是表面摩擦系数为
	$$C_f = \frac{\tau_w}{\frac{1}{2}\rho U^2}$$ $$= 3\frac{\mu U}{\delta}\frac{1}{\rho U^2}$$ $$= 3\frac{\sqrt{Re_x}}{4.64}\frac{1}{Re_x}$$ $$= \frac{0.647}{\sqrt{Re_x}}$$

6.9 Turbulent Boundary Layer for Incompressible Flow Along a Flat Plate

6.9 沿平板不可压缩流动湍流边界层

Vast majority of boundary layers encountered in practice are turbulent over most of their length. Thus, turbulent boundary layer analysis is regarded as of greater fundamental importance than the laminar boundary layer. But there is no exact theory available for turbulent flat plate flow, although there are many elegant computer solutions of the boundary layer equations, using various empirical models for turbulent eddy viscosity. The most widely accepted result is simply an integral analysis similar to our study of laminar profile approximation [Equation (6.9)].

The momentum equation [Equation (6.8)] may be applied to the turbulent boundary layer, as no limiting assumptions were made in its derivation. However, a new relation for the velocity profile, up through the boundary layer, will have to be found and the shear stress will no longer be obtained simply from the

工程实际中遇到的绝大多数边界层在大部分长度上都是湍流的。因此，湍流边界层的分析比层流边界层更基础更重要。但是，对于湍流平板流动，尽管有许多利用各种湍流涡黏性经验模型求解的完美的边界层方程计算机解，却没有完全精确的适用理论。而最普遍采用的结果只是一个类似于层流速度分布估算［式（6.9）］的积分分析。

由于推导过程中没有做任何假设，动量方程［式（6.8）］也可以应用于湍流边界层。然而，必须找到一个新的边界层法向速度分布关系，而剪切应力也不再只是流体黏度和流速分布

product of fluid viscosity and the gradient of the velocity profile. Because of the basic similarity between the development of boundary layer within circular pipes and over flat plates, Prandtl suggested that the results from pipe can be applied to the analysis of flat plate turbulent boundary layers. We know that, the boundary layer growth in pipes is limited to the pipe radius R, so that $u = U$ at $r = R$, and the mean velocity in turbulent pipe flow is known to be about $0.8U$. The velocity distribution in such flow is adequately represented by the Prandtl power law,

$$\frac{u}{U} = \left(\frac{y}{\delta}\right)^n \tag{6.35}$$

where $n = \frac{1}{7}$ for $Re_x < 10^7$. Obviously, this profile breaks down at the wall, where $y = 0$. But the presence of laminar sublayer makes the velocity to decrease linearly to zero at the wall, and this profile being tangential to the power law.

To develop the analogy between flat plate and pipe flow, it is necessary to appreciate that $\delta = R$, in the fully developed region, and to develop some relations for τ_w to replace which no longer applies.

$$\tau_w = \mu \frac{\partial u}{\partial y}$$

Blasius proposed that, for smooth pipes, the shear stress at the wall could be expressed by

$$\tau_w = f \frac{1}{2} \rho \bar{u}^2 \tag{6.36}$$

where $\bar{u}$ is the mean velocity of the flow, which is equal to $0.8U$ and f is an empirical constant known as friction factor, which is a function of flow Reynolds number, based on pipe diameter d, and the ratio of wall roughness to pipe diameter. Thus,

梯度的乘积。由于圆管内部边界层的发展与平板上边界层发展规律基本相似，普朗特提出可把圆管中结果应用于平板湍流边界层分析。我们知道管道中边界层的增长至多到管径 R，于是 $r = R$ 处 $u = U$，而且已知湍流管道流中平均速度为 $0.8U$，这一流动中速度分布足以用普朗特幂次分布规律表示为

式中，$Re_x < 10^7$ 时 $n = \frac{1}{7}$。显然，在壁面（$y = 0$）处，这一速度分布不成立。层流底层的存在使得流速线性降为壁面处的零，且这一分布曲线与幂次分布规律相切。

为了建立平板流动和管道流动之间的相似，应在充分发展区令 $\delta = R$，然后建立一些 τ_w 的关系式以替代不再适用的公式

布拉休斯提出，对于光滑管，壁面上的剪切应力可表示为

式中，$\bar{u}$ 是管道平均流速，大小为 $0.8U$；f 是一个称为摩擦系数的经验常数，它是一个由管道直径 d 计算的雷诺数以及壁面粗糙度和管道直径之比的函数。于是

$$\tau_w = f \frac{1}{2}\rho(0.8U)^2$$

| and, as Blasius developed the expression | 并且，由于布拉休斯建立了表达式 |

$$f = \frac{0.079}{Re^{1/4}} = \frac{0.079}{(\rho \bar{u} d/\mu)^{1/4}}$$

| Thus, for smooth pipes, we have | 因此，对于光滑管，有 |

$$\tau_w = \frac{1}{2}\rho(0.8U)^2 0.079[\mu/(\rho 0.8U 2R)]^{1/4}$$

| If $\delta = R$, this becomes | 如果 $\delta = R$，上式变为 |

$$\tau_w = 0.0225\rho U^2 (\mu/\rho U \delta)^{1/4} \qquad (6.37)$$

| As the assumption of zero pressure gradient has been made, Equation (6.8) can be applied. Thus, | 由于假设压力梯度为零，可利用式（6.8），于是 |

$$\tau_w = \rho U^2 \frac{d\delta}{dx} \int_0^\delta \frac{u}{U}\left(1 - \frac{u}{U}\right) d\eta$$

| is | 即 |

$$\tau_w = \rho U^2 \frac{d\delta}{dx} \int_0^1 (1-\eta^{1/7})\eta^{1/7} d\eta \qquad (6.38)$$

| where $u/U = (y/\delta)^{1/7} = \eta^{1/7}$. Therefore, | 式中，$u/U = (y/\delta)^{1/7} = \eta^{1/7}$，因此 |

$$\tau_w = \frac{7}{72}\rho U^2 \frac{d\delta}{dx} \qquad (6.39)$$

| Equating Equations (6.37) and (6.39), we get | 联立式（6.37）和式（6.39），得到 |

$$\delta^{1/4} d\delta = 0.2314(\mu/\rho U)^{1/4} dx$$

| Integrating this, we get | 积分后得到 |

$$\frac{4}{5}\delta^{5/4} = 0.2314(\mu/\rho U)^{1/4} x + c$$

| Now, assuming the boundary layer to be turbulent right from the leading edge of the plate, which is reasonable if the plate is long compared to the length of the laminar boundary layer, then $\delta = 0$ at $x = 0$ and $c = 0$. Hence, | 如果与层流边界层长度相比平板是较长的，则 $x=0$ 处 $\delta=0$，$c=0$。那么现在可以假设边界层从平板前缘开始即为湍流，因此 |

$$\delta^{5/4} = 0.29(\mu/\rho U)^{1/4} x$$

| or | 或 |

$$\delta = \frac{0.37x}{Re_x^{1/5}} \qquad (6.40)$$

Thus, the thickness, δ of a turbulent boundary layer increases as $x^{4/5}$, which is far faster than the laminar boundary layer thickness which increases as $x^{1/2}$. The shear stress on the flat surface may be determined by eliminating δ between Equations (6.37) and (6.40). Thus, the shear stress becomes	于是，湍流边界层厚度 δ 以 $x^{4/5}$ 增长，远比层流边界层厚度以 $x^{1/2}$ 增长要快很多。 平板壁面的剪切应力可通过联立式（6.37）与式（6.40）消掉 δ 后得出。于是，剪切应力变为

$$\tau_w = 0.029\rho U^2 (\mu/\rho Ux)^{1/5}$$

and the skin-friction force acting on the wall becomes	又有作用在壁面上的表面摩擦力为

$$F = \int_0^l \tau_w \mathrm{d}x \quad (\text{per unit width, 单位宽度})$$

where l is the length of the plate. Now, substituting for τ_w, we get	式中，l 为平板长度。现在，代入 τ_w 后得到

$$\begin{aligned}F &= \int_0^l 0.029\rho U^2(\mu/\rho Ux)^{1/5}\mathrm{d}x \\ &= \left[0.029\rho U^2(\mu/\rho U)^{1/5}\frac{x^{-4/5}}{4/5}\right]_0^l \\ &= 0.036\rho U^2 l(\mu/\rho Ul)^{1/5}\end{aligned}$$

The skin-friction coefficient is	表面摩擦系数为

$$C_f = \frac{F}{\frac{1}{2}\rho U^2 l} \quad (\text{per unit width, 单位宽度})$$

that is,	即

$$C_f = \frac{0.072}{Re_l^{1/5}} \qquad (6.41)$$

This expression is valid for Reynolds number up to 10^7, but experimental results indicate that a better approximation is given by	此式对于高达 10^7 的雷诺数才成立，但实验结果表明更为准确的近似为

$$C_f = \frac{0.074}{Re_l^{1/5}} \qquad (6.42)$$

Prandtl has suggested subtracting the length of the laminar layer, resulting in an expression	普朗特提出从式（6.42）中减掉层流边界层的长度，可得到表达式

$$C_f = \frac{0.074}{Re_l^{1/5}} - \frac{1700}{Re_l}$$

This is valid for Re from 5×10^5 to 10^7.	此式适用于 $5\times(10^5 \sim 10^7)$ 的雷

To extend the Reynolds number range further, Schlichting employed the logarithmic velocity distribution, for pipe under turbulent flow condition, resulting in a semi-empirical relation

$$C_f = \frac{0.455}{(\log_{10} Re_l)^{2.58}} \tag{6.43}$$

Equation (6.43) can be used for Reynolds number greater than 10^7. For $Re < 10^7$, Equation (6.43) gives values for C_f that are very close to those given by Equation (6.42), and consequently engineers commonly use Equation (6.43) over the entire range of Reynolds number above 500000. Comparing Equations (6.41) and (6.26b), it is seen that the skin-friction is proportional to the $\frac{9}{5}$ power of velocity of the main stream and the $\frac{4}{5}$ power of plate length for the turbulent boundary layer, compared to $\frac{3}{2}$ and $\frac{1}{2}$ powers, respectively, for the laminar boundary layer.

6.10 Flows With Pressure Gradient

In flows with zero pressure gradients, for example, flat plate flow, the point of inflection (PI) is at the wall itself. In flows with adverse pressure gradient, a point of inflection occurs in the boundary layer and its distance from the wall increases with the strength of the adverse pressure gradient.

Examine the flow through a convergent-divergent passage shown in Figure 6.14. The flow profile of Figure 6.14, usually occurs in sequence as the boundary layer progresses along the wall of a body.

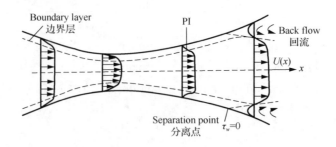

Figure 6.14　Viscous flow through a convergent-divergent duct

图 6.14　流过收缩-扩张管道的黏性流

The nozzle flow is a flow with favorable gradient and never separates nor does the throat flow where the pressure gradient is approximately zero. But in diffusers velocity decreases and pressure increases, an adverse gradient is created. When the diffuser angle is too large and the adverse pressure gradient is excessive, the boundary layer will separate at one or both walls, with back flow, increased losses, and poor pressure recovery. In diffuser literature this condition is called diffuser stall. This is usually referred to as boundary layer separation. At this stage we should note that the boundary layer theory can compute only up to the separation point and after which it is invalid.

收缩管部分是具有正压梯度的流动，不会分离，压力梯度几乎为零的喉管流动也是一样。但在扩张管中，由于流速降低、压力增加，产生了逆压梯度。当扩张角足够大时，逆压梯度过大，在一侧或两侧壁面就会出现边界层分离，伴随着回流、增大的能量损耗以及很小的压力恢复。在研究扩压器的文献中这种情况被称为扩压器失速。这通常被归结为边界层分离。此处我们应注意，边界层理论只适用于边界层分离点前的计算，分离点之后边界层理论不适用。

6.11 Laminar Integral Theory

6.11 层流积分理论

Both laminar and turbulent theories can be developed from Karman's general two-dimensional boundary layer integral relation, which can be expressed in terms of $U(x)$ as

层流或湍流边界层理论都可以由卡门的通用二维边界层积分关系建立，用 $U(x)$ 表示为

$$\frac{\tau_w}{\rho U^2} = \frac{1}{2}C_f = \frac{\mathrm{d}\theta}{\mathrm{d}x} + (2+H)\frac{\theta}{U}\frac{\mathrm{d}U}{\mathrm{d}x} \qquad (6.44)$$

where $\theta(x)$ is the momentum thickness and $H(x) = \delta^*(x)/\theta(x)$ is the shape factor. Also, we know that negative $\mathrm{d}U/\mathrm{d}x$ is equivalent to positive $\mathrm{d}p/\mathrm{d}x$, i.e. an adverse pressure gradient. We can

式中，$\theta(x)$ 是动量厚度；$H(x) = \delta^*(x)/\theta(x)$ 是形状因子。而且，我们已知负的 $\mathrm{d}U/\mathrm{d}x$ 等于正的 $\mathrm{d}p/\mathrm{d}x$，也就

integrate Equation (6.44) to determine $\theta(x)$ for a given $U(x)$, if we correlate C_f and H with momentum thickness. This has been done by examining typical velocity profiles of laminar and turbulent boundary layer flows for various pressure gradients.

Consider the velocity profiles shown in Figure 6.15, illustrating that the shape factor H is a good indicator of pressure gradient.

The higher the H the stronger the adverse pressure gradient, and separation occurs approximately at

$$H \approx 3.5 \quad \text{for laminar flow}$$

and

$$H \approx 2.4 \quad \text{for turbulent flow}$$

Laminar profile exhibits the S-shape and a point of inflection (PI) with an adverse gradient. But in the turbulent profile the point of inflection is typically buried deep within the thin viscous sublayer.

是逆压梯度。如果我们用动量厚度建立 C_f 和 H 的关系，可以就一个给定的 $U(x)$ 对式(6.44)积分来计算 $\theta(x)$。这在不同压力梯度下，层流与湍流边界层流动的典型速度分布分析中已经介绍了。

观察图 6.15 中所示的速度分布，可以看出形状因子 H 能够很好地反映压力梯度的大小。

H 越高，逆压梯度越强，分离则大致发生在

$H \approx 3.5$（层流流动）

和

$H \approx 2.4$（湍流流动）

层流速度分布呈 S 形且有一个逆压梯度的转折点（PI）。但在湍流分布中，转折点通常位于很薄的黏性底层中。

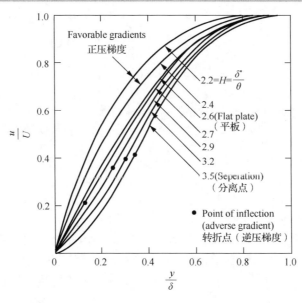

Figure 6.15 Velocity profiled with pressure gradient for laminar flow

图 6.15 层流中速度分布与压力梯度的关系

For laminar flow, a simple and effective method | 对于层流流动，思韦茨提

was developed by Thwaites, who found that Equation (6.44) can be correlated by a simple dimensionless momentum thickness variable λ, defined as

$$\lambda = \frac{\theta^2}{\nu}\frac{dU}{dx} \tag{6.45}$$

Using a straight-line fit to his correlation, Thwaites was able to integrate Equation (6.44) in a closed form, to result in

$$\theta^2 = \theta_0^2 + \frac{0.45\nu}{U^6}\int_0^x U^5 dx \tag{6.46}$$

where θ_0 is the momentum thickness at $x=0$ (usually taken to be zero). Separation ($C_f = 0$) was found to occur at a particular value of λ. The value is

$$\lambda = -0.09 \tag{6.47}$$

Finally, Thwaites correlated values of the dimensionless shear $S = \frac{\tau_w \theta}{\rho U}$, with λ, and his graphed result can be curve-fit as follows.

$$S(\lambda) = \frac{\tau_w \theta}{\rho U} \approx (\lambda + 0.009)^{0.62} \tag{6.48}$$

This parameter is related to skin friction coefficient by the identity

$$S = \frac{1}{2}C_f Re_\theta \tag{6.49}$$

Equations (6.46) to (6.49) constitute a complete theory for the laminar boundary layer with variable $U(x)$, with an accuracy of ±10 % compared with exact digital computer solution of the laminar boundary layer equations [Equation (6.21)].

Example 6.3

Water flows at a speed of 1m/s over a flat plate of length 1m in the flow direction. The boundary layer is tripped to make it turbulent at the leading edge. Assuming 1/7 power turbulent velocity profile, find the boundary layer thickness, displacement thickness, and the wall shear stress at the trailing edge of the plate. Solve the same problem if the flow over the plate is laminar. The kinematic viscosity of water is $10^{-6} \text{m}^2/\text{s}$.

次幂（η 的 1/7 次幂），计算边界层厚度、位移厚度以及平板后缘处的壁面剪切应力。如果绕平板流动是层流，求解同样的问题。水的运动黏度为 $10^{-6} \text{m}^2/\text{s}$。

Solution

As we know, $U = 1\text{m/s}$, $L = 1\text{m}$, $\nu = 10^{-6}\text{m}^2/\text{s}$, at the trailing of the plate, $x = L = 1$ the Reynolds number is

解

已知 $U = 1\text{m/s}$，$L = 1\text{m}$，$\nu = 10^{-6}\text{m}^2/\text{s}$。在平板尾部，$x = L = 1$，雷诺数为

$$Re_L = \frac{UL}{\nu} = \frac{1 \times 1}{10^{-6}} = 10^6$$

This is greater than the flat plate critical Reynolds number of 5×10^5, hence the flow is turbulent. Therefore, by Equation (6.40), we have the boundary layer thickness as

此值要高于平板流动的临界雷诺数 5×10^5，因此流动是湍流的。所以，由式（6.40），得到边界层厚度为

$$\delta \approx \frac{0.37L}{(Re_L)^{1/5}} = \frac{0.37 \times 1}{(10^{-6})^{1/5}} = 0.0233\text{m} = 23.3\text{mm}$$

For $u/U = (y/\delta)^{1/7} = \eta^{1/7}$, the displacement thickness, by Equation (6.2), is

因为 $u/U = (y/\delta)^{1/7} = \eta^{1/7}$，由式（6.2），边界层位移厚度为

$$\begin{aligned}\delta^* &= \int_0^\delta (1-\eta^{1/7})\text{d}y \\ &= \left[\eta - \frac{\eta^{1/7+1}}{1/7+1}\right]_0^\delta \\ &= \left[\eta - \frac{7}{8}\eta^{8/7}\right]_0^\delta \\ &= \delta - \frac{7}{8}\delta^{\frac{8}{7}} \\ &\approx \frac{\delta}{8} \\ &= 23.3/8 \\ &= 2.91\text{mm}\end{aligned}$$

The wall shear stress is given by

壁面剪切应力为

$$\begin{aligned}\tau_w &= 0.029\rho U^2(\mu/\rho UL)^{1/5} \\ &= 0.029 \times 10^3 \times 1^2 (10^{-6})^{1/5} \\ &= 1.83\text{Pa}\end{aligned}$$

For Laminar flow, by Equation (6.25), we have

对于层流流动，由式（6.25），有

$$\delta_L = \frac{5}{\sqrt{Re_L}} \times L$$

$$= \frac{5}{\sqrt{10^6}}$$

$$= 5\text{mm}$$

By Equation (6.33), we have	由式（6.33），有

$$\frac{\theta}{L} = \frac{0.664}{\sqrt{Re_L}}$$

By Equation (6.34), we have	由式（6.34），有

$$\delta^* = 2.59\theta$$

$$= \frac{2.59 \times 0.664}{\sqrt{Re_L}}$$

$$= \frac{2.59 \times 0.664}{\sqrt{10^6}}$$

$$= 1.72\text{mm}$$

By Equation (6.26a), we have	由式（6.26a），有

$$C_f = \frac{0.664}{\sqrt{Re_L}}$$

Therefore,	因此

$$\tau_w = C_f \frac{1}{2}\rho U^2$$

$$= \frac{0.664}{\sqrt{10^6}} \times \frac{1}{2} \times 10^3 \times 1^2 = 0.332\text{Pa}$$

Example 6.4

(a) Determine the frictional drag acting on one side of a smooth flat plate of length 0.5m and width 0.15m, placed longitudinally in an air stream of velocity 1m/s, at sea level. (b) Find the boundary layer thickness and the shear stress at the trailing edge of the plate.

Solution

(a) At standard sea level the atmospheric temperature is 15℃(= 288K). Therefore, the viscosity of air becomes

例 6.4

（a）一长 0.5m、宽 0.15m 的光滑平板沿长边放置于海平面上流速为 1m/s 的气流中，求平板一侧表面上受到的摩擦阻力。（b）求边界层厚度以及平板末端的剪切应力。

解

（a）在标准海平面上大气温度为 15℃（=288K），因此空气黏度为

$$\mu = (1.46 \times 10^{-6}) \times \frac{288^{3/2}}{288+111}$$

$$= 17.88 \times 10^{-6} \text{kg}/(\text{m}\cdot\text{s})$$

The air density is	空气密度为

$$\rho = \frac{p}{RT}$$
$$= \frac{101325}{287.26 \times 288}$$
$$= 1.225 \text{kg} / \text{m}^3$$

The Reynolds number at the plate end is	平板末端的雷诺数为

$$Re = \frac{\rho UL}{\mu}$$
$$= \frac{1.225 \times 1 \times 0.5}{17.88 \times 10^{-6}}$$
$$= 34256$$

This is less than 500000, hence the flow is laminar. Therefore, the skin friction coefficient, by Equation (6.26), becomes	此值小于 500000，因此流动为层流流动。于是由式（6.26），表面摩擦力系数变为

$$C_f = \frac{0.664}{\sqrt{Re}}$$
$$= \frac{0.664}{\sqrt{34256}}$$
$$= 0.00359$$

The frictional drag acting on one side of the plate becomes	作用在平板一侧的摩擦阻力变为

$$F_f = \frac{1}{2}\rho U^2 \frac{A}{2} C_f$$
$$= 0.5 \times 1.225 \times 1^2 \times \frac{0.5 \times 0.15}{2} \times 0.00359$$
$$= 8.25 \times 10^{-5} \text{N}$$

where A is the surface area of the plate. (b) By Equation (6.25), the boundary layer thickness at the trailing edge of the plate is	式中，A 为平板的表面积。 （b）由式（6.25），平板末端的边界层厚度为

$$\delta = \frac{5L}{\sqrt{Re_L}}$$
$$= \frac{5 \times 0.15}{\sqrt{34256}}$$
$$= 4.05 \text{mm}$$

The shear stress at the plate end, by Equation (6.28b), is	根据式（6.28b）平板末端的剪切应力为

$$\tau_w = 0.332\sqrt{\frac{\rho\mu}{L}}U^{3/2}$$

$$= 0.332\sqrt{\frac{\rho UL}{\mu}\frac{\mu U}{L}}$$

$$= 0.332\sqrt{Re}\frac{\mu U}{L}$$

$$= 0.332\sqrt{34284}\times\frac{17.88\times 10^{-6}\times 1}{0.5}$$

$$= 0.0022\text{Pa}$$

Example 6.5 　　A submarine has been approximated to a rectangular box of length 360m, width 70m and height 25m. If it is traveling at a speed of 24km/h, in seawater with kinematic viscosity $\nu = 1.4\times 10^{-6}\text{m}^2/\text{s}$ and density $1020\text{kg}/\text{m}^3$, determine the skin friction drag of the submarine and the power required to overcome the drag. **Solution** 　　The Reynolds number of the flow over the submarine, based on its length is	**例 6.5** 　　一个潜水艇被近似为一个长360m、宽70m、高25m的长方体，如果以24km/h的速度在运动黏度$\nu = 1.4\times 10^{-6}$ m²/s、密度为1020kg/m³的海水中行驶。计算艇体的表面摩擦阻力以及克服阻力所需的功率。 **解** 　　根据艇长计算绕艇体流动的雷诺数为

$$Re_L = \frac{UL}{\nu}$$

$$= \frac{(24/3.6)\times 360}{1.4\times 10^{-6}}$$

$$= 1.714\times 10^9$$

Considering each surface of the submarine to be a flat plate, we have the skin friction coefficient by Equation (6.43) as	假设潜水艇的各个表面均为平板，由式（6.43）可得表面摩擦阻力系数为

$$C_f = \frac{0.455}{(\log_{10}Re_L)^{2.58}}$$

$$= \frac{0.455}{[\lg(1.714\times 10^9)]^{2.58}}$$

$$= 0.00147$$

The skin friction drag is given by	表面摩擦阻力为

$$D_f = C_f A_s \frac{1}{2}\rho U^2$$

where A_s is the surface area. For the given vessel, we have	式中，A_s 为表面积，由给定的艇体有

$$A_s = 2[(360 \times 70) + (360 \times 25)]$$
$$= 68400 \text{m}^2$$

| Note that the surface areas of the face and base of the box are neglected here. Thus, | 注意,此处忽略了艇体的首部与尾部的表面积,因此 |

$$D_f = 0.00147 \times 68400 \times \frac{1}{2} \times 1020 \times (24/3.6)^2$$
$$= 2.279 \text{MN}$$

| The power required to overcome this drag is | 克服阻力所需的功率则为 |

$$P = DU = 2.279 \times (24/3.6)$$
$$= 15.19 \text{MW}$$

Example 6.6

A viscous fluid flows over a flat plate such that the boundary layer thickness at a distance 1.3m from the leading edge is 12mm. Assuming the flow to be laminar, determine the boundary layer thickness at a distance of (a) 0.2m, (b) 2.0 m, and (c) 20 m from the leading edge.

Solution

For laminar boundary layer, from Equation (6.25), the boundary layer thickness is

例 6.6

绕平板的黏性流动中,距前缘 1.3m 处的边界层厚度为 12mm。假设流动是层流的,计算距前缘（a）0.2m、（b）2m 与（c）20m 处的边界层厚度。

解

对于层流边界层,由公式（6.25）,边界层厚度为

$$\delta = \frac{5x}{\sqrt{Re_x}}$$

| where x is the distance from the leading edge. Therefore, the Reynolds number at $x = 1.3$ m is | 式中,x 表示距离平板前缘的距离。因此在 $x = 1.3$ m 处的雷诺数为 |

$$Re_x = \frac{5^2 x^2}{\delta^2} = \frac{5^2 \times 1.3^2}{0.012^2} = 293403$$

| that is, | 即 |

$$\frac{\rho U 1.3}{\mu} = 293403$$

| (a) Assuming ρ, U and μ to remain constant, at $x = 0.2$m, we have | （a）假设 ρ、U 和 μ 都是常数,在 $x = 0.2$m 处则有 |

$$Re_{0.2} = 293403 \times \frac{0.2}{1.3} = 45139$$

| Therefore, the boundary layer thickness at $x = 0.2$m becomes | 因此,在 $x = 0.2$m 处边界层厚度为 |

$$\delta_{0.2} = \frac{5 \times 0.2}{\sqrt{45139}} = 4.71 \text{mm}$$

(b) At $x = 2$m, the Reynolds number is	（b）在 $x = 2$m 处，雷诺数为

$$Re_2 = 293403 \times \frac{2}{1.3} = 451389$$

Therefore,	于是，

$$\delta_2 = \frac{5 \times 2}{\sqrt{451389}} = 14.88 \text{mm}$$

(c) At $x = 20$m, the Reynolds number is	（c）在 $x = 20$m 处，雷诺数为

$$Re_{20} = 293403 \times \frac{20}{1.3} = 4513892$$

Therefore,	因此

$$\delta_{20} = \frac{5 \times 20}{\sqrt{4513892}} = 47.07 \text{mm}$$

6.12 Summary

In this chapter, what we gave is just a glimpse of boundary layer theory rather than in-depth information, to get an idea about this important aspect of fluid flow analysis. The concept of boundary layer, the calculation of displacement thickness, momentum thickness and kinetic energy thickness, the types of boundary layer and the solutions of boundary layer were talked about. Several boundary layer problems of interest were also discussed by giving practical examples.

6.13 Exercises

Problem 6.1 Air flows over a flat plate. At a given location along the plate the boundary layer thickness is $\delta = 50$mm. At this location, what would be the boundary layer thickness if it were defined as a distance from the plate where the velocity is 97 percent of the freestream velocity rather than 99 percent? Assume the flow to be laminar.

6.12 小结

本章只给出了关于边界层理论的一些粗浅的信息，而非深入的内容，以使读者对流体流动分析中这一重要部分有个初步的概念。因此，介绍了边界层概念，位移厚度、动量厚度和动能厚度的计算，边界层类型和边界层的求解，并通过实例讨论了几个普遍关注的边界层问题。

6.13 习题

题 6.1 空气流过一个平板。在沿平板方向某一给定位置处边界层厚度为 $\delta = 50$mm。假设流动是层流的，问：如果边界层厚度定义为边界层中速度是来流速度的 97% 处而不是 99% 处与平板壁面的距离，则此处的边界层厚度应为多少？

Problem 6.2 Consider an incompressible boundary layer with freestream velocity U_∞ = constant, as shown in Figure 6.16. Assuming a velocity distribution

$$u = U_\infty \left(2\frac{y}{\delta} - \frac{y^2}{\delta^2} \right)$$

determine the wall shear stress $\tau_0(x)$ and the boundary layer thickness $\delta(x)$. The given parabolic velocity profile is an approximation to a laminar boundary layer profile.

[Ans: $\tau_0 = 0.365\rho U_\infty^2 \sqrt{\dfrac{\mu}{\rho U_\infty x}}$, $\delta = \dfrac{5.484x}{\sqrt{Re_x}}$]

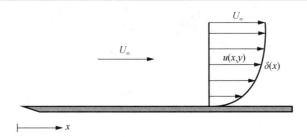

Figure 6.16 Boundary layer on a flat plate

图 6.16 平板边界层

Problem 6.3 A laminar boundary layer formed on one side of a flat plate of length l produces a drag D. How much must the plate be shortened if the drag on the new plate has to be $D/4$? Assume that the freestream velocity remains the same. Explain your answer physically.

[Ans: $l/16$]

Problem 6.4 A hut to serve as temporary housing near a seashore has to be designed. The hut may be considered as a closed (without leak) semi-cylinder of radius 5m, mounted on tie-down blocks, as shown in Figure 6.17. The viscous effects are neglected and the flow over the top of the hut is identical to the flow over a cylinder for $0 \leq \theta \leq \pi$. When we calculate the flow over the upper surface of the hut, the presence of the air space under the hut is neglected. The air under the hut is at rest and the pressure is equal to the stagnation pressure p_t. What are the net lift and drag forces per unit depth of the hut? The wind speed is 50m/s and the stagnation freestream properties are those of standard sea-level condition. Also, find the lift and drag coefficients.

[Ans: 40833.33N, 0N, 2.67, 0]

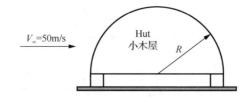

Figure 6.17 A semi-cylindrical hut in a seashore

Problem 6.5 Calculate the skin friction coefficient C_f for flow over the top surface of a flat plate at a Reynolds number 2.5×10^5, if the flow is (a) entirely laminar and (b) turbulent, assuming the power law $u/U = (y/\delta)^{1/7}$.

[Ans: (a) 0.00265; (b) 0.006]

Problem 6.6 Water at a speed of 0.5m/s flows over a flat plate of length 6m. Assuming the boundary layer to be laminar, determine the boundary layer thickness and the wall shear stress at the (a) centre and (b) trailing edge of the plate. Take the viscosity of water as 1.12×10^{-3} kg/(m·s).

[Ans: (a) 0.0013m, 0.0718Pa; (b) 0.0183m, 0.0508Pa]

Problem 6.7 A thin flat plate of length 0.3m and width 1m is installed in a water tunnel as a splitter. The freestream speed is 1.6m/s and the boundary layer on the plate is laminar, with a parabolic velocity profile. Neglecting the pressure drag, determine the total viscous drag force acting on the plate.

[Ans: 1.6N]

Problem 6.8 A thin flat plate of length $L = 0.3$m and width $b = 1$m is installed in a water tunnel as a splitter. The freestream speed is 2m/s, at 20℃, and the velocity profile is approximated as

$$\frac{u}{U} = 2\left(\frac{y}{\delta}\right) - \left(\frac{y}{\delta}\right)^2$$

and the boundary layer thickness over the plate at this condition is given by $\delta / x = 5.48 / \sqrt{Re_x}$. Show that, the total drag force on one side of the plate can be expressed as $D = \rho U^2 \theta_L b$. Calculate (a) the momentum thickness θ_L and (b) the drag D.

[Ans: (a) 0.283mm; (b) 1.132N]

Problem 6.9 An aircraft of wing span 12m and average chord 2m flies at a speed of 200km/h at an altitude where the air density is 1.2kg/m³ and kinematic viscosity is 1.5×10^{-5}m²/s. Assuming that the wing skin friction is the same as that on a smooth flat plate of same dimensions, calculate (a) the frictional drag and (b) the power required to overcome this frictional drag.

[Ans: (a) 280.3N; (b) 15.57kW]

References
参考文献

[1] Schlichting H. Boundary Layer Theory[M]. New York: McGraw-Hill, 1955.